PERISHABILITY FATIGUE

CRITICAL LIFE STUDIES

CRITICAL LIFE STUDIES

JAMI WEINSTEIN, CLAIRE COLEBROOK, AND MYRA J. HIRD, SERIES EDITORS

The core concept of critical life studies strikes at the heart of the dilemma that contemporary critical theory has been circling around: namely, the negotiation of the human, its residues, a priori configurations, the persistence of humanism in structures of thought, and the figure of life as a constitutive focus for ethical, political, ontological, and epistemological questions. Despite attempts to move quickly through humanism (and organicism) to more adequate theoretical concepts, such haste has impeded the analysis of how the humanist concept of life is preconfigured or immanent to the supposedly new conceptual leap. The Critical Life Studies series thus aims to destabilize critical theory's central figure, life—no longer should we rely upon it as the horizon of all constitutive meaning but instead begin with life as the problematic of critical theory and its reconceptualization as the condition of possibility for thought. By reframing the notion of life critically—outside the orbit and primacy of the human and subversive to its organic forms—the series aims to foster a more expansive, less parochial engagement with critical theory.

Luce Irigaray and Michael Marder, *Through Vegetal Being: Two Philosophical Perspectives* (2016)

Jami Weinstein and Claire Colebrook, eds., *Posthumous Life: Theorizing Beyond the Posthuman* (2017)

Penelope Deutscher, *Foucault's Futures: A Critique of Reproductive Reason* (2017)

PERISHABILITY FATIGUE

FORAYS INTO ENVIRONMENTAL LOSS AND DECAY

VINCENT BRUYERE

Columbia University Press   *New York*

Columbia University Press
Publishers Since 1893
New York   Chichester, West Sussex
cup.columbia.edu
Copyright © 2018 Columbia University Press

Cataloging-in-Publication Data available from the Library of Congress
ISBN 978-0-231-18858-6 (cloth)
ISBN 978-0-231-18859-3 (pbk.)
ISBN 978-0-231-54794-9 (ebook)

Cover design: Julia Kushnirsky
Cover art: Copyright © Massimo Brega

CONTENTS

PREFACE

MYRRHA'S PRAYER

L atin poet Ovid recounts Myrrha's story in the tenth book of his *Metamorphoses*. After nine wandering months and pregnant with her father's child, she finally comes to a stop both "tired of living and scared of dying."[1] She can't take it anymore: "O, if there are any gods who hear my prayer, I do not plead against my well deserved punishment, but lest, by being, I offend the living, or, by dying, offend the dead, banish me from both realms, and change me, and deny me life and death!" In response to her words of exhaustion, Myrrha receives from a benevolent divinity a surrogate body to both bear her child and her punishment to term:

> While she was still speaking, the soil covered her shins; roots, breaking from her toes, spread sideways, supporting a tall trunk; her bones strengthened, and in the midst of the remaining marrow, the blood became sap; her arms became long branches; her fingers, twigs; her skin, solid bark. And now the growing tree had drawn together over her ponderous belly, buried her breasts, and was beginning to encase her neck: she could not bear the wait, and she sank down against the wood, to meet it, and plunged her face into the bark.

And so Myrrha lives on, but as an incense tree (*commiphora myrrha*) weeping warm tears of aromatic sap over the world and her newborn child.

Like its fallen and metamorphosed creatures, Ovid's text has many lives and survives in many forms. In *Ovide Moralisé* the Christian medieval tradition managed to turn Myrrha's life around by painting her plight in surprising virtuous terms.[2] In *La Métamorphose d'Ovide figurée*, published by Jean de Tournes, engraver Bernard Salomon gave her in the middle of the sixteenth century a visual flesh to be reborn to, as a feminine silhouette inhabiting a vegetal form. If today I return to Myrrha's transgressive, transspecies, and transformative trajectory, it is to focus on the remains of her impossible demand: "change me, and deny me life and death." Today it is as if these words had found a new coherence in the fact that processes of decay and death exist as domains of intervention and contestation rather than solely as the existential condition of being mortal and transient. Being perishable is to be targeted by protocols of conservation, prediction, and termination. It is to find oneself waiting, having to wait, and yet unable "to bear the wait" for a projected future to take its statistical toll and for a negative prognosis to claim its existential cull. In the context of cancer culture, it is to "live in prognosis," understood as constitutive of what it means to be ill.[3] In the context of palliative care, being perishable is to find oneself dying and having to take responsibility for it, to make a choice regarding the next course of action, to face a set of options that will always appear too limited.

Perishability mirrors the concept of viability introduced by French historian of biology Georges Canguilhem: "The historical succession of organisms, on the basis of what is now known as prebiotic chemical evolution, is a succession of claimants not powerful enough to be living beings that are more than merely

FIGURE O.I Bernard Salomon, *La Métamorphose d'Ovide figurée* (Lyon, 1557). Courtesy of Albert and Shirley Small Special Collections Library, University of Virginia.

viable—that is to say, beings that are fit to live, but lack any guarantee of succeeding totally in doing so."[4] If viability accounts for the fact of having no more than a shot at life rather than a right to it (as in a sovereign right to decide life and death or as "the right of the social body to ensure, maintain, or develop its life"),[5] perishability then calls the shots and counts the rounds. With that in mind, Myrrha is less a prototype of biopolitical heroism, reclaiming the terms of her living death in the history of power over life, than a figure of the "claimant": she is not in a position to choose nor to grant, only to receive, endure, and be transformed ("change me"). There is almost a utopian ring to her expression of a desire to be neither dead nor alive and rest somewhere in the middle between storage and burial. But if the specifics of her metamorphosis unsettle a history of power written by the governance of loss and decay, as power to store or to bury, to conserve, to prognosticate, or to terminate, it is only provisionally. It is only as a thought experiment that nevertheless lives on as an untimely reminder that managing perishability is more than a matter of getting better at freezing genetic diversity in remote storage units (chapter 1). It is not just a matter of getting better at keeping decay at bay (chapter 2), containing toxicity in eternal burial grounds (chapter 3), fostering cellular immortality (chapter 4), screening for premalignant neoplasms (chapter 5), or sedating pain in terminal patients (chapter 6). It is also a matter of understanding in what capacity the governance of loss and decay defines the biographical borders of an exhausting present. No matter what the current field of biotech possibilities already is or soon will be able to achieve, the scenario Ovid leaves us with has no part to play in the experts' talk devising what, in the face of environmental and human degradation, responsibility, accountability, and feasibility ought to look like. Holding on to the memory of Myrrha's exhaustion and prayer, my main concern in *Perishability Fatigue* is precisely to make room for positions of nonexpertise in the domain of operations carved out by seed banks, bioengineering proposals, genetically modified organisms, toxic waste repositories, tissue culture, cancer narratives, and palliative care. It is to bring documents familiar to policy analysis such as protocols of conservation, prediction, and termination to a form of unfamiliar scrutiny. It is to revisit the boundaries being assigned to a present that, on the one hand, imagines de-extinction programs pursuing the genetic resuscitation of charismatic species—wooly mammoth, Tasmanian tiger, and passenger pigeon—and that, on the other hand, designs protocols of "compassionate disposal" reimagining the lifespan of frozen embryos left from in vitro fertilization.

Developed in a context where thousands of frozen embryos have accumulated in storage waiting for decisions, policies, and even incidents to happen, protocols of compassionate disposal describe, in effect, a new form of burial practice that consists in transferring spare frozen embryos to "the patient's uterus at a time in her monthly cycle when she is very unlikely to become pregnant."[6] A 2009 survey assessing couples' indecisions about whether or not to donate their spare embryos to other infertile couples or opt for compassionate disposal transports us into the world of an eighteenth-century novel where bastards reclaim their birthrights and honest bourgeois wake up in a kinship nightmare: "Many parents envisioned scenarios in which their children might unwittingly fall in love with a full sibling, or complications would arise in the distribution of inheritance, or they would answer the doorbell one day 'to face an angry teenager demanding an explanation about why he was given away.'"[7] The logic that these parents deploy is preemptive, but there is more to it. They abide by scenarios that manage to reconnect, even if unwittingly, pasts, presents, and futures held separate in vats of liquid nitrogen.

At the other end of the spectrum, prospects of de-extinction technology thrive on the intermediary state they too posit and project. For environmental scientist Stuart Pimm, de-extinction is a fantasy in which "there is nothing involving the real-world realities of habitat destruction, of the inherent conflict between growing human populations and wildlife survival. Why worry about endangered species? We can simply keep their DNA and put them back in the wild later."[8] In the meantime, the "wild" is gone, or at least it is gone as a habitat to become a "before" with no "after." The de-extinction fantasy, or experiments in de-extinction, depending on the perspective adopted, might live up to its promises or "fray," to use Lauren Berlant's vocabulary— only time will tell.[9] But in the meantime, and as a story some people tell, rehearse, rewrite, or act upon, the genetic resurrection of extinct species does not fail to designate a transitional moment nested between the history of science and technology and a cultural memory of transience and exhaustion. For sociologist Karin Knorr Cetina, this sense of the meantime corresponds to what she describes as the rise of a culture of life grounded in the promises of biomedical science to perfect, enhance, and extend life, by opposition to the Enlightenment definition of a culture of the human grounded in social salvation. It is not clear, in the outline of the narrative she puts forward, what happens during the period of transition, as a culture of life rises and the culture of the human loses its grip on the lives it used to qualify. She writes, for instance, that "sociality is likely to be a permanent feature of human life. But the forms of sociality are changing, and the regions of social structuring may expand or contract in conjunction with historical developments."[10] *Perishability Fatigue* takes stock of a number of possibilities, proposals, projects, protocols, prognoses, and prototypes that diagnose, envision, and negotiate this transitional moment as simultaneously yet to be determined and yet in an apparent state of disarray. My objective is not to gather a scattering of broken promises but to assess the relational templates and forms of continuity in time that fulfilled biotech and biomed promises afford to some and refuse to others.

The task is akin to isolating layers in the overall accumulation of narratives, practices, and procedures in which the belief in a principle of sustainability is anchored—belief in the sense that, as literary scholar Karen Pinkus puts it in conversation with Cameron Tonkinwise, Australian specialist of design research, to be sustainable, that is, "'to enjoy resources in the present' one would have to know what 'the present' is; and to believe that consumption in that 'present' is calculable, measurable and acceptable."[11] The accounting effort that has to be put into this is quite formidable, magnetizing enough to produce forms of being independent (sustainability through self-sufficiency) or, conversely, of being collective (sustainability through service design and share usage),[12] but also forms of inquiry into the making and unmaking of forms of continuity in time. As such, it does make sense for Tonkinwise to ask in a review of Allan Stoekl of *Bataille's Peak: Energy, Religion, and Postsustainability* (2007): "What form . . . will future earth-dwellers, forevermore exposed to their own finitude by the loss of ready-to-hand energy, take?" to then suggest that, "perhaps, given the variabilities and limits of their alternative energy sources, they will be much more historically sensitive sets of humans."[13] The fatigue in "perishability fatigue" is on the side of this historical sensitivity.

If sustainability science and disaster preparedness flesh out mitigation models and survivalist scenarios to face resource depletion and species extinction, they also generate opportunities for critical interventions into what it means and takes to be perishable. For Gillen D'Arcy Wood, sustainability studies represent a new generation of humanistic intervention defined by

xiv € PREFACE: MYRRHA'S PRAYER

its engagement with complex system literacy rather than by a Luddite or elegiac rejection of scientific culture. Its ambitious mandate is to offer "rich answers to central questions: for instance, how have human instincts, desires, and aspirations fed historically into larger ideological construction of nature, be they providential, instrumental, or sustainable."[14] In that spirit, *Perishability Fatigue* takes the reader on an associative argumentative path traced through interdisciplinary premises. The effect of disorientation that may ensue is deliberate. It has the value and the virtues of a simulation experience that aims at bringing into light the need for a critical literacy of the meantime committed to imagining points of intersection and passage between the vertiginous temporal perspectives offered by climate modeling and biobanking and narratives of stranded existence collected by scholars in the fields of critical race theory and disability studies. That being said, the provisional position the book attempts to secure in the context of sustainability studies and biomedical culture does not pledge access to a lived experience of perishability. Rather, it anthologizes scenes, scripts, and scenarios of being perishable, which, in the wake of Myrrha's prayer, take responsibility for the dreams of conservation, preservation, and regeneration that technologies of storage and burial are expected to fulfill; and redraw the dividing lines between past, present, and future, between diagnosis and prognosis, between storage and burial, between "the needs of the present" and "the ability of future generations to meet their own needs,"[15] between a form of degradation one can claim responsibility for (like pollution as a modality of leaving traces and marking a territory)[16] and a form of degradation that, in the case of climate change, terminal illness, and systemic violence, exceeds or at least challenges the structures of responsibility in place.

The strength of the overall edifice lies in the lines each chapter draws between the withholding structures that characterize the governance of loss, storage, and decay and pressing demands to bear with loss and the defacing possibility of radical change whose vocabulary and tropes I foraged—hence the "forays"—in Ovid, Rabelais, Montaigne, Kafka, Michel Foucault, Jean-Dominique Bauby, Roland Barthes, and Rebecca Skloot, in early modern visual culture, or in the folktale of Bluebeard. As lines are being drawn, forays commemorate another ephemeral assemblage preserved in the archives of psychology. In *Playing and Reality*, British psychoanalyst Donald Winnicott recounts the case of a seven-year-old boy presenting with what seems to be a developmental disorder. His parents brought him to the Psychology Department of the Paddington Green Children's Hospital in March 1955:

> They said that the boy had become obsessed with everything to do with string, and in fact whenever they went into a room they were liable to find that he had joined together chairs and tables; and they might find a cushion, for instance, with a string joining it to the fireplace. They said that the boy's preoccupation with string was gradually developing a new feature, one that had worried them instead of causing them ordinary concern. He had recently tied a string round his sister's neck (the sister whose birth provided the first separation of this boy from his mother).[17]

Winnicott is able to get him to stop playing with strings, but in a note added ten years after his first clinical observations we learn that the boy grew up to become a housebound drug addict. Perhaps, in a different place and a different time, he would have cooked meth, using his binding skills to turn innocuous domestic products into the speed needed to meet the demands that late capitalist conditions of production put on bodies.[18] In a different place and a different time, pulling different strings, drawing different lines, he may have grown up to become a geographer or even an architect redefining exiguous habitable volumes and attending to precarious conditions of coexistence.[19] In the world where the boy has grown up to become wary of holding structures to the point of challenging a domestic assemblage of things and persons, "thirty minutes were all that separated the quotidian from annihilation."[20] In this world the disjuncture between offensive capabilities and protective measures defines the situation in which the term *megadeath* became available to political strategists to designate the instantaneous wiping out of millions of lives by a nuclear attack.[21] In these circumstances, the boy's noncompliance with the order of things has the value of a quasi-cosmological intervention destined to secure a set of unraveling relations—no matter how tentative the gesture, no matter how frail the shielding structure, no matter how provisional the woven installation. In the context of this book, the string that joins together visual and literary objects, built structures, and biographies is a figure of nonexpertise. It has been repurposed as the thread tying a perishable present together.

In the first chapter, I turn toward the founding principles of the Svalbard Global Seed Vault project, an Arctic archive designed to preserve the crop capital of humanity, as well as toward a 2012 bioethics proposal addressing climate change remediation from the perspective of incentivized human engineering. The point is not to determine whether these ventures will be able to fulfill their promises one day. Rather, it is to examine what it is that a seed bank wishes into existence and, in more general terms, how future lives and biographies are wished into existence by technologies of perishability. In chapter 2 I return to the Svalbard project, but this time through the angle of visual culture and, more specifically, through the notion of "still life" as a category in critical discourse. Why? To create a parallel between two entombment effects, that is, two ways of removing

objects from circulation and two ways of managing, acknowl-
edging, and even harnessing perishability: by storing germ-
plasms in a vault or by isolating objects in a painted frame. If
Svalbard's connection with crop history and the governance of
genetic diversity seems pretty straightforward, representational
objects such as Dutch still life paintings are not commonly associated with biotechnology and food security. However, I argue
that this type of connection becomes valid and visible in a situation where GMO technologies engineer fruits that do not rot.

In chapter 3 the experience of continuity in time corresponds
in equal parts to a future entombed, as in Svalbard, and an exercise in managing perishability, as in still life painting. But here
we are no longer talking about seeds or crops. Being managed is,
on the one hand, the slow decay over millennia of radioactive
waste stored/buried in the desert of New Mexico under the
guardianship of the U.S. Department of Energy, and, on the
other hand, the radio-vulnerability of human life. The governance of perishability defines postnuclear eternity as a form of
meantime and organizes the meantime as a zone of operation to
be occupied through the development of security protocols and,
incidentally, in one of the security proposals solicited by the U.S.
Department of Energy, through the recoding of literary culture.
While chapter 3 focuses on how a provocative protocol of nuclear
waste management reinvests the cultural memory of cautionary
tales, chapter 4 focuses on how another form of management of
decay, tissue culture, affects the process of biographical capture
in Rebecca Skloot's best seller *The Immortal Life of Henrietta
Lacks* (2011). In *The Immortal Life of Henrietta Lacks* (2011),
biographical capture results from the clash between the history
of a woman of color and the history of her body, between the
record of her death from cervical cancer in 1951 at Johns Hopkins
Hospital in Baltimore and the culture of her mutated cells in

petri dishes across the scientific world. Here the word *capture* and its attendant effect of erasure cannot be heard without historical overtones of racial violence, dispossession, and deprivation. With the next chapter, the inquiry into the culture of perishability moves from the afterlife of Henrietta's terminal cancer to the status of diagnostic and prognostic categories in contemporary cancer culture. The chapter stands out in its structure. It is composed of a series of six biographical improvisations—or oncoscripts—exploring what it means and what it takes for six patients, real, fictional, or hypothetical, to occupy the meantime charted by oncology.

It seems only fitting to return to Myrrha's metamorphosis on my way toward the last chapter. As she "loses her former senses with her body" (v. 499), Myrrha also loses the voice in which she had articulated her prayer. With this loss, it is her metamorphosis that is lost to experience or, perhaps, that would have been lost without the benevolence of the poetic archive that recites her prayer to perishability. It is through this lens that I propose to read the account Jean Dominique Bauby gave of his paraplegic and aphasic life in *The Diving Bell and the Butterfly* (1997). Unlike Myrrha's, Bauby's immobilization is not premeditated. It is the consequence of a cardiovascular accident that left him radically transformed, speechless and motionless, save for a blinking eyelid that he used to transmit his memoir. The exemplarity of his text in a culture of perishability is undeniable. It touches on the intensification of the relation of coimplication that exists between the history of what it will have meant to have and to lose one's body and the history of how that bodily state and its logic of dispossessions has gained access to discourse in the palliative present.

ACKNOWLEDGMENTS

As this project comes to a close, or at least a pause, it is an immense pleasure to acknowledge the institutions that supported the research behind the book and the friends, colleagues, and students whose benevolence nudged the manuscript toward completion. *Perishability Fatigue* owes much of its inception to a visiting fellowship at the Australian National University in 2012. I am grateful to the director of the Humanities Research Center, Debjani Ganguly, for such a rewarding intellectual experience in such a beautiful setting. My stay in Canberra was made even more memorable by the camaraderie of Peter Gratton and Karen Pinkus and, unexpectedly, by ANU's harpsichord collection. Karen was the first to bring to my attention Svalbard Seed Vault during one of our conversational hikes through the National Botanic Gardens. Her pioneering work on climate change, her incisive wit and bounteous creativity have remained for me a constant, emboldening source of inspiration in my own writing. In its earlier stage, the project also received the support of the Gimon Collection at Stanford and of the Center for Global Studies at Penn State. In State College I had the good fortune to be adopted by the Science,

Technology, and Society Program. I am grateful to Susan
Squier for introducing me to feminist science studies and
encouraging me to explore intersections between health and
humanities. I am grateful for the friendship and thoughtful
counsel of Jennifer Boittin, Heather McCoy, Bénédicte Moni-
cat, and Allan Stoekl during my time at Penn State. Special
thanks go to Jon Abel, Atia Satar, and Jens-Uwe Guettel for
sharing their knowledge of horror cinema—Jon also read the
manuscript at a crucial stage and offered his much-needed
enthusiasm and insights. I want to thank Carrie Jackson for
opening her home during my summer retreats in central Penn-
sylvania, and the South Gill neighborhood for its neighborly
spirit.

The book owes a lot to Emory vibrant community of scholars. I want to express my sincere appreciation to the collegiality
of Deepika Bahri, Geoffrey Bennington, Munia Bhaumik,
Sarah Blanton, Rosemarie Garland-Thomson, Jonathan Gold-
berg, Lynne Huffer, Dalia Judovitz, Michelle Lampl, Valerie
Loichot, Elissa Marder, Michael Moon, Claire Nouvet, Ben
Reiss, Nathan Suhr-Sytsma, Elizabeth Wilson, and Subha
Xavier. Very special thanks are due to Elissa Marder. Her guid-
ance in all writing matters and her encouragements have been
instrumental to the development of this book as well as to my
own academic development. I also thank the members of my
graduate and undergraduate seminars on the history of the body
for giving me the opportunity to try out ideas and associations,
and Iman Williams in particular for reminding me to read
Hortense Spillers again. I would be remiss if I did not mention
Amandine Ballart and Leslie Church for watching over the
administrative details of our little world. The Emory Center for
Digital Scholarship team at the Robert W. Woodruff Library
and Melanie Kowalski have been precious allies when it came

to gathering illustrations and chasing sources. The Emory College of Art and Science was most supportive, providing leave time in the last phase of the book's completion.

I am deeply grateful to my editor Wendy Lochner for believing in such a wild and untamed project, to the readers whose comments transformed my manuscript, and to Claire Colebrook, Myra Hird, and Jami Weinstein for including *Perishability Fatigue* in their superb Critical Life Studies series. Lowell Frye and Susan Pensak were both invaluable in seeing through the project in its production and copyediting stages.

It would have been impossible to stay the course without the memories of summer days spent with my parents, my sisters, Margaux and Fanny, her partner Yohan, my nieces, Bertille and Emma, and my extended family. For many years now, Maureen, Luke, Oak, Juniper, and their pets have been more than generous in food, stories, and humanimal warmth. At home, Sumter provides his own unconditional share of emotional support and wagging love. Finally, I offer this strange *ex-voto* of a book, modeled over time after my own likeness, to Aaron in recognition of the joy he brings to the present and as a tribute to our future together.

An earlier version of the argument developed in chapter 1 appeared in *diacritics* 41, no. 3 (2013): 60–79, copyright © 2014 Cornell University. A section of chapter 3 was previously published in "Of a Survivalist Tone Adopted in Literary Studies," *symplokē* 22, nos. 1–2, Austerity (2014): 167–79. Part of chapter 4 originally appeared in *Mosaic, an interdisciplinary critical journal* 48, no. 4 (2015): 123–36.

PERISHABILITY
FATIGUE

1

BEING FABULOUS AS THE
CLIMATE CHANGES

And just see how the World has gone to seed.
—Rabelais, *The Third Book*

Much as a nineteenth-century architecture of pleasure channeled Parisian wealth into spaces made
for the exhibition, consumption, and circulation of
goods, a twenty-first-century architecture of finitude channels
narratives of sustainability into vaulted offerings to an uncertain future. This architecture finds its epitome in the creation
of the Svalbard Global Seed Vault—also known as the
"Doomsday Vault." Inaugurated in 2008, this storage facility
can accommodate 4.5 million seed samples divided in three
chambers encased deep inside a mountain situated on an arctic
island belonging to Norway, 810 miles from the North Pole.[1] Its
entrance, blasted out of the permafrost, is situated 130 meters
above sea level and is therefore safe from the rise of the oceans.
It is on a politically stable jurisdiction, anchored in a geologi-
cally stable terrain, and placed within a consistently cold environment even in the event that the cooling system fails. More
than a place, the vault is a placeholder. In media representations,

the Svalbard project functions as a geopolitical locus in a global-
ized fantasy of sustainability in which it is possible to pass on a
desire for futurity. Along with other community-oriented
cooperative biobanking projects on the edge of biocapitalism
(such as Kenny Ausubel's Seeds of Change project), it locates a
promise of regeneration through biodiversity.[2] Svalbard vault
briefly found a counterpart in the Seed Cathedral, the now dis-
mantled UK Pavilion at the 2010 Shanghai Expo. A creation of
Thomas Heatherwick's studio, the Seed Cathedral consists of "a
box, 15 metres high and 10 metres tall. From every surface pro-
trude silvery hairs, consisting of 60,000 identical rods of clear
acrylic, 7.5 metres long, which extend through the walls of the
box and lift it into the air. Inside the pavilion, the geometry of
the rods forms a space described by a curvaceous undulating
surface. There are 250,000 seeds cast into the glassy tips of all the
hairs."[3] It is a counterpart, not only because Kew's Millennium
Seed Bank provided the encased seeds but also because Heatherwick's structure was as ephemeral as the Svalbard vault is eternal, as transparent and aerial as the other is opaque and telluric.
Or perhaps a counterpart, in the way that thousands of seeds set
in thousands of transparent stems simultaneously staged and
interrupted a potential of germination.

On the occasion of the one-year anniversary of the Svalbard
vault, Norwegian Minister of Agriculture and Food Lars Peder
Brekk cautioned, "no one knows if the Seed Vault ever will be
needed. No one knows if and when the seeds will be sent back
to their depositor to restore a seed collection that has been lost.
That's how insurance policies work."[4] Svalbard thus turns the
present into a time of indefinite prescription in which a foretold
doomsday event is both hypothetical and yet real enough. As
such, the vault calls for an event, an accident, or, by increments,

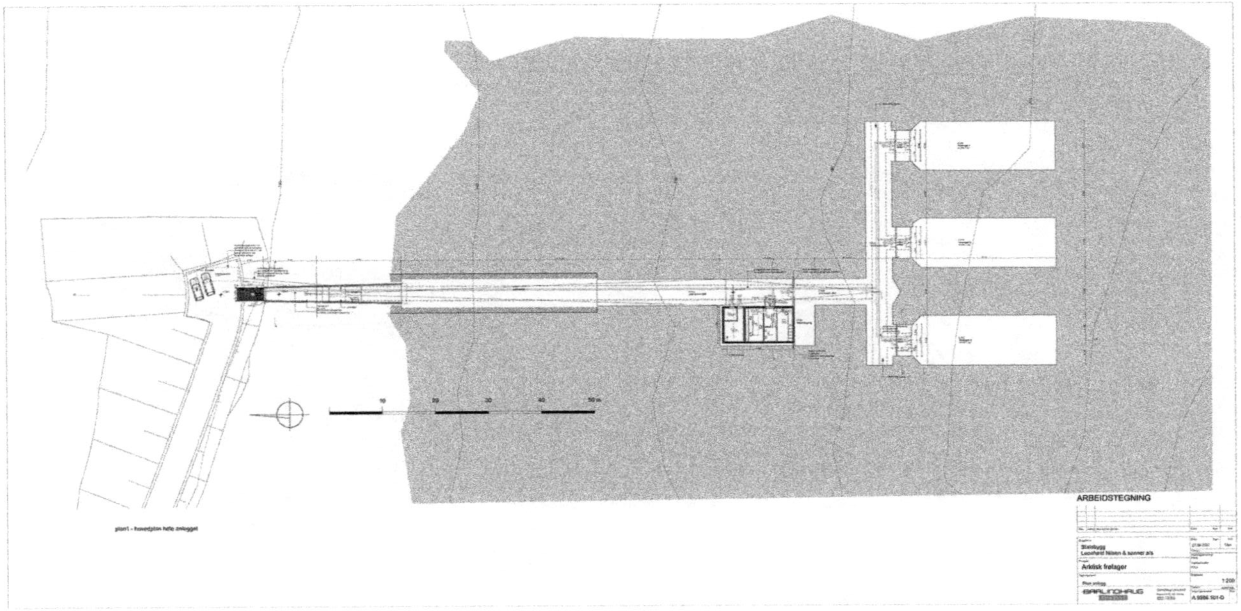

FIGURE 1.1 Svalbard Global Seed Vault. Drawing by the Directorate of Public Construction and Property, Norway.

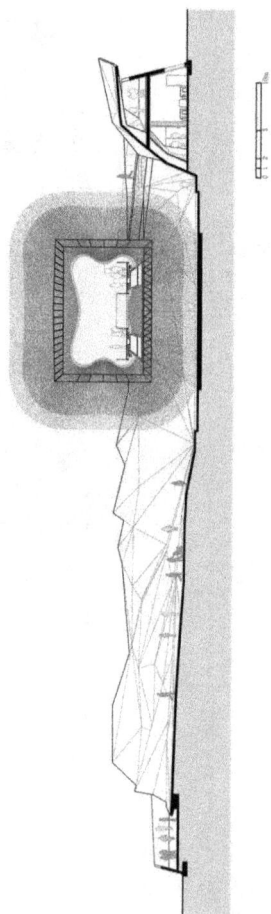

FIGURE 1.2 UK Pavilion, 2010. Reproduced with permission. Heatherwick Studio.

for a critical threshold to be met. If it could speak, as in the realm of fables and in the words of the Book of the Apocalypse, the vault would say: Come/*Viens*/*erkhou*.[5] In this light, storing is as much an eschatological and moral issue, as it is a technical and technological endeavor. The seed bank ensures continuity by summoning a future against which it has been built. It marks and signifies the passage from a past to a future, from a time of climate felicity correlated with the birth of agriculture in the Holocene to a time of climatic disruption, mass extinction, and biodiversity loss. It defines a present in which doing something about the future consists in investing in burial grounds, storage facilities, and other repositories. Here futurity is both a domain of intervention and an element in the syntax of moral obligation to a world in the making that leaves behind, but without necessary achieving a heritage status, a trail of experiments and costly prototypes designed to withstand survivalist aspirations. Svalbard entombs potential annihilation to keep it from annihilating potentialities. With Karen Pinkus we can say that this model of sustainability functions as a gift of time, a gift to a future whose very politico-actuarial existence depends on the availability of what is given up now through storage or burial. In Svalbard the future feeds on resources that will have been saved up or put on hold somewhere in the Arctic Circle.[6] In the meantime and in turn, the politics of ecological and economic adjustment assemble a present of constant assessments that both reflects and curtails the possibility of radical change—for Svalbard also seals off possible scenarios and questions.[7] It precludes other projects that might have entailed radical changes in the ethics of bioengineering and current agendas in crop politics, or even in the way societies respond to risk assessments. Here disaster futurity is the regulatory instance in the circulation of what-if scenarios that project a future available to forms of governance and technological interventions.

In the Brundtland Commission Report *Our Common Future* (1987), the regulatory instance is developmental, or, as encapsulated in its most often quoted proposition: "Sustainable development is development that meets the needs of the present without compromising the ability of future generations to meet their own needs." Regulation is assured by the determination of external costs that would be necessary to take into account in order to effectively balance off energy input and output, offset environmental impacts, and achieve a certain level of sustainability. It is hard however, Allan Stoekl remarks, to put a price tag on externalities: "True sustainability would involve the pricing of resources in such a way that their finitude is reflected in their prices (and thus the reduction of consumption of resources to acceptable levels). But how does one represent, through price, the fact that a given resource will, at some point, no longer be available? To whom, when?"[8] Similarly, a question of representation remains pending, opened by the Svalbard's temporal horizon: from where is it possible to address a vaulted politics of conservation and sustainability? Is it from the present that devised a specific holding? Is it the present that the holding is projecting—the long-awaited moment of the opening, or the moment when it is time to discard a hold that became useless, perhaps unsustainable? Or is it the present that was avoided or postponed from the get-go and thereby sealed off by the storage/burial device itself? In each case, the point of reference is internal to the plan of action adopted. Svalbard and its representations coexist in a continuum that is hard to navigate. To punctuate this continuum, I turn to an unlikely source, an early modern fable of cryopreservation found in *The Fourth Book*

(1552) by François Rabelais. One of the most flamboyant figures of French humanism, Rabelais's name evokes a world of unbridled abundance, irreverent excess, and spirited banter between well-read giants and their gaudy entourage. Nothing could be farther removed from the ethos of frugal sustainability.

As they sail on the edge of the glacial sea, Pantagruel and his companions encounter disembodied sounds and atmospheric voices: "Mates, do you hear anything? It seems to me I hear people talking up in the air [*parlans en l'air*], but all the same I don't see anyone. Listen!"[9] *Paroles en l'air* are literally "suspended words," but also, in modern French, "words spoken in vain." Whatever these *paroles en l'air* translate as, they are up for grabs and interpretation. Pantagruel's erudition strikes first by anchoring the aerial phenomenon in the history of philosophy and turning the encounter into a metaphysical event:

> I've read that a philosopher named Petron was of the opinion that there were several worlds touching one another in the form of an equilateral triangle, at the base and center of which he says were the abode of Truth and the habitat of Words, Ideas, the exemplars and images of all things past and future, and all around these was the Age. And in certain years, at long intervals, part of these fell upon humans like cattarrhs, and as the dew fell upon Gideon's fleece; part of them remain reserved for the future, until the consummation of the Age. I also remember that Aristotle says that Homer's words are prancing, flying, moving, and consequently animate. Moreover, Antiphanes used to say that Plato's doctrine was like certain words, which in some region or other, in the depths of a hard winter, freeze and turn to ice in the cold of the air, and are not heard.[10]

The skipper later gives a more terrestrial explanation to the frozen soundscape. The floating words are the acoustic remainders of battle noises once trapped by the cold into enigmatically colorful words. Now thawing in the vernal warmth and warm hands ("quelque peu eschauffez entre nos mains"),[11] they release a mix of onomatopoeia and words:

> hin, hin, hin, hin, his, ticque, torche, lorgne, brededin, brededac, frr, frrr, frrr, bou, bou, bou, bou, bou, bou, bou, traccc, trac, trr, trr, trr, trrr, trrrr. On, on, on, on, ououourouon: goth, magoth, and I know not what other barbarous words; [the captain] said that these were vocables from the crashing and from the neighing of the horses at the time of the clash. Then we heard some other, coarse ones, and in unfreezing they made a noise, some as of drums and fifes, some as bugles and trumpets.[12]

Among many puzzled readers of these famous passages, Renaissance scholar Marion Leathers-Kuntz suggests that the episode of the frozen words reconstitute allegorically an anecdote found in Guillaume Postel (1510–1581). Postel recounts a Paris winter so cold that ink froze in inkpots, forcing him to breath onto his pen to write a book—*De Orbis Concordia*—whose publication the Sorbonne's censors eventually did not allow.[13] A philological reconstitution that anchors the whereabouts of a text in climate history is most suggestive for someone interested in food security and disaster preparedness. However, the passage in *The Fourth Book* does not leave us with a question—"what are the frozen words?"—but with two contradictory gestures and a paradoxical lesson in sustainability. The narrator-chronicler explains: "I wanted to preserve a few of the gay quips [*motz de gueule*] in oil, the way you keep snow and ice, and then to wrap them up in clean straw. But Pantagruel refused, saying that it was folly to store up things which one is never short of, and which are always plentiful, as gay quips [*motz de gueule*] are among good and jovial Pantagruelists."[14] On the one hand, adding to the record he is keeping, he attempts to preserve some of the frozen words in their frozen state and postpone their explosive utterance. On the other hand, intervening in the travel log process, Pantagruel enjoins his chronicler to let the frozen words thaw on the ground of their perpetual availability in such spirited company.

There is something compelling about a fable that envisioned, centuries before Edison's phonograph, the possibility to record defunct voices, but also, following German media theorist Friedrich Kittler, the possibility to store time outside texts and music scores, in ice, only to celebrate in the end their destruction by thawing.[15] Philology, by contrast, runs on an ethics of conservation. It opts for the preservation of the frozen words through the reconstitution of what they are, or could be—and as it is, *motz de gueule* can be many things that pertain to their color ("gules" in heraldry), to their tone, their content, and the fact they are spoken. It makes sure that the sense of the words is not completely lost, or that it is possible to formulate hypotheses regarding what they really are. In this particular case, philology seems to be missing the fact that the "frozen words"—the fabulous thing described in Rabelais's text and the episode itself as parable or fable—belong to the future and belong to it under modalities that are not understood in terms of preservation. If there is one, the lesson of the frozen words is not given as the direct expression of someone's authority (the interpreter's or the dead's), but in the name of a certain confidence in a form of continuity in time. At this juncture, Rabelais's fable of cryopreservation leaves us with two ways of understanding what a resource is: in one case, something to be preserved and, in the other, something to be expended. Even more significantly, the targeted resource (the *paroles en l'air*) to be conserved or expended appears to be interpretive and is commemorated as such by the text itself as at once frozen and thawing, at once material (a frozen by-product of a particular episode in climate history, perhaps a manifestation of the Little Ice Age) and allegorical (words subject to censorship, living dead words in print). The fable of the frozen words powers interpretation as a fuel, not as source of energy. I borrow the distinction from Pinkus's creative engagement with the timelines of climate change remediation and the search for alternative energy sources. The fable fuels interpretation because of its relation to potentiality: some of the frozen words could be preserved or be left to thaw. Nonpreservation is an option envisioned and explored by the text.[16]

All concrete and steel, there is nothing fabulous or interpretative about Svalbard. There is no room for ambiguity regarding the nature of what the seed vault contains, sets apart, and preserves against time—the time of biodiversity loss and climate change projections. However, the intermediate state that Svalbard posits (as insurance and as "a call for the event") is neither the temporality of crop history nor that of climate futures. It is its own thing, a form of continuity in time internal to the structure itself that does not preexist the structure—something experts in climate change remediation and risk assessment usually do not talk about. Something, I contend, to talk about by reference to Rabelais's frozen words. Of course Rabelais is not the answer to the problem Svalbard has been designed to solve and stage, even if in its structure "let it thaw" could be an answer. The fable is only a reminder that Svalbard is a proposal, a fable too. That is, it commemorates Svalbard as a form-giving event with respect to the picture of a future defined by freezing and thawing affordances. From this unusual angle, projects, proposals, and policies monitoring resources and making recommendations on forms of continuity in time take on a quasi-fabulous identity. They demand and deserve a quasi-philological mode of attention.

The 2012 human engineering proposal penned by Matthew Liao, Anders Sandberg, and Rebecca Roache is quite demanding in this regard. Published in the journal *Ethics, Policy, and Environment*, "Human Engineering and Climate Change" explores biotechnological alternatives to programs seeking market-regulated behavioral change to address global warming, as well as alternatives to equally controversial geoengineering programs such as solar radiation management (SRM). If geoengineering is too risky, reforming deeply ingrained modes of consumption is an uphill battle too tedious to be fought and won in time—for the clock is ticking. The authors' suggestions range from a pharmacological patch inducing meat intolerance to help individuals become vegetarian, and therefore participate actively in the reduction of livestock farming, to oxytocin treatment enhancing altruism and empathy—precious qualities indeed in times of resource scarcity—to more ad-hoc radical genetic modifications. These measures, insist the authors, would not be forced upon a population but suggested, through tax breaks or sponsored healthcare incentives. The ideas are meant to be provocative. Almost apologetically, Liao, Sandberg, and Roache write:

> We are well aware that our proposal to encourage having smaller, but environmentally-friendlier human beings is *prima facie* outlandish, and we have made no attempt to avoid provoking this response. There is a good reason for this, namely, we wish to highlight that examining intuitively absurd or apparently drastic

ideas can be an important learning experience, and that failing to do so could result in our missing out on opportunities to address important, often urgent, issues.[17]

The outlandish is not spoken in vain, for fables, no matter how far-fetched they sound, can be turned into resources. Here, too, conservation meets interpretation. By the time the provocation has subdued, the future will have turned Liao, Sandberg, and Roache's *paroles en l'air* into something that can be exchanged against time, more time: a chance to survive a history that is yet to come.

For a historian of medicine and public health, the measures lined up in the human engineering proposal have a past. They belong to the corpus of modern panoplies designed to "take apart and repair, cut off, replace, take out, add, correct, or straighten" wayward bodies.[18] The waywardness may have a new name and a new context, but the body to be reformed and outfitted has not changed. In fact, it just got stronger, its fiction more believable. The rather intriguing meat patch idea is not far from more venerable and established "orthopedic" procedures from the Enlightenment period. In one unforgettable cosmological *vanitas*, which appeared for the first time in *Le Postmoderne expliqué aux enfants*—a year before the Brundtland Report on sustainable development—Jean-François Lyotard indirectly resets the history of this reformative effort in a timeline that culminates with the forthcoming explosion of the sun in about four and half billion of years. Meanwhile, humankind is getting ready for a general evacuation: "The reason for the number of experiments challenging the human body lies in the necessity for the human body to be made either adaptable to or commutable with another body, another device, more adaptable to extraterrestrial conditions, and it seems to me that this horizon has to be thought in the general perspective of techno-scientific development, insofar as this development is aimed at the emigration of humankind from the earth."[19] For the time being, before the vaporization of ontology and planet Earth altogether, humankind believes in endings and endorses exterminating angels along with redeeming figures. It comes up with the notion of sustainability and releases the Brundtland Report along the way. It gets better at governing itself and managing perishability. It tells stories, including that story of the sun's ending, stretching to its conjectural limit the difference between realism and reality, between history and historiography: "What a Human and his/her Brain—or rather the Brain and its Human—would resemble at the moment when they leave the planet forever, before its destruction; that the story does not say."[20] (What humankind might resemble at the moment when it needs the seed vault the story does not say.) That, in fact, the story cannot say, for it is itself one of the contraptions challenging the survivability of the species. The Liao, Sandberg, and Roache proposal is a contraption describing contraptions targeting and remodeling the shape of *bios* in bioethics, biopolitics, and biotechnology. The transformational potential that Liao, Sandberg, and Roache envision takes place in a time in which, much like in Lyotard's biotechnological escalation fable, there is no fundamental difference between a slow (too slow, the proposal insists) self-formation process that cultivates human potentiality through the arts and humanities and the biochemical fast-tracking of behavioral change for the sake of the survival of humankind.[21] The outlandish is not spoken in vain. The fabulous will come to fruition.

As with Svalbard and Rabelais's *paroles en l'air*, Liao, Sandberg, and Roache's proposal posits a form of continuity in time that is neither that of crop history nor that of climate futures. It projects a temporality in which to derive one's existence as responsible existence. If the machinery cogs imagined by Liao, Sandberg, and Roache are new and unusual, its cruelty principle is not. In Latin, *cruor* is the blood that has been shed in a sacrificial context, by opposition to *sanguis,* the blood that courses in the veins. *Sanguis* is the liquid of life that flows freely in opposition to *cruor*—the blood of a life made available to the law and before which life ought to respond and, in some cases, pay with death and in death the price of its responsibility. It is the logic that Franz Kafka describes in his short story, "In the Penal Colony" (1914).

Like the article "Human Engineering and Climate Change," Kafka's text is a contraption describing a contraption and the political system it informs: a sentencing mechanism designed to write the violated law on the violator's body with a harrow. Because the mode of inscription is so intricate and cuts so deep, the victim perishes in the process, without trial and without even knowing the accusation. A fair trial would be unnecessary, explains the officer serving the machine: "It would be useless to give [the victim] that information. He experiences it on his own body." But in Kafka's penal colony, as noted by Lyotard in his reading of the short story, the machine is already old and left in disrepair.[22] It has bad press in the new political order and does not gather a community around its bloody proceedings anymore. There is another judicial machine in town, a parliamentary machine that sustains deliberations and mandates fair trials. By the end of the fable, the old machine collapses on its own, killing its most devoted servant. The destruction of the machine puts an end to an era. It opens onto an emancipatory future of civil rights. Liao, Sandberg, and Roache's human engineering proposal is futuristic in its science-fictional allure, but it does not belong to the future prepared by the collapse of the old penal machine. Quite the opposite. Its release signals the return of a certain form of exacting cruelty in a normative world that only knows procedures and technical rationality. With human engineering, the communal spectacle that embodies obligation, responsibility, prescription, and survivalist citizenship is more arresting and pronounced than ever, even if it is bloodless. The proposal is a machine that does not require much blood (*cruor*)—it is a cold machine, like Svalbard. It operates at the level of the blood stream (*sanguis*) without interrupting it, or at the molecular level: oxytocin will flow with *sanguis*, and preimplantation genetic diagnoses (PGD) aimed at selecting shorter children (to lower a population's metabolic rate and thus reduce its energy needs) will happen upstream. No cuts, no incisions are necessary, only a more or less gentle culling or, for the enthusiasts, incentives to cultivate new possibilities. Cruelty is redirected toward a timeless future always already in the process of being reclaimed by a technonormative apparatus of reasoned sacrifice.

There might be something Kafkaesque in Liao, Sandberg, and Roache's proposal, but there is also something definitely Ovidian in its hypothetical birthing of a responsible humanity eager to reduce its unmanageable carbon footprint. Yes, tree planting, after all, is a form of carbon sequestration and climate change mitigation, but this is not what I have in mind with Myrrha's arborescence. If her metamorphosis is not an elective procedure, it remains a creative solution offered to an impossible demand of continuity in time even after the loss of bodily integrity and even after the loss of sentience ("Though she has lost her former senses with her body, she still weeps").[23] In Liao, Sandberg, and Roache's proposal, continuity in time is contingent upon shape-shifting and form-giving opportunities that would make humans shorter, frugal, altruistic, and capable of seeing in the dark (so as to reduce their reliance on electric bulbs). To think of human engineering in terms of metamorphosis (rather than eugenics) is to approach the future of biography from a different angle: not just as the record of a lifespan, but as the active and incentivized shaping of a timeline. (At the end of "In the Penal Colony" the Condemned Man survives the destruction of the machine set up to make him die. His release has the unexpected result not of letting him live but of making him live.)[24] It is to think continuity in time not as a given but as something opened to intervention, and even as a domain of operation in itself. Conversely, to recode a set of material-technical relations geared toward the actuarial future through a cultural memory of perishability is a way to sidestep the machinery of the proposal to focus on its more or less successful attempts at articulating what it means for a body to have a future, that is, to exist in time as the climate changes.

2

STILL LIFE WITH GENETICALLY
MODIFIED TOMATO

*What need have I of the lemon's principial form? What my
quite empirical humanity needs is a lemon ready for use, half-
peeled, half-sliced, half-lemon, half-juice, caught at the precious moment it exchanges the scandal of its perfect and useless
ellipse for the first of its economic qualities, astringency.*
—Roland Barthes, "The World as Object"

eveloped in 1989 under the name Flavr Savr by
Calgene, an agrobiotech company based in Davis,
California, CGN-89564-2 was the first genetically
modified organism approved by the Food and Drug Administration (FDA). Flavr Savr had a longer shelf life than a regular
tomato thanks to its "antisense PG" gene reducing, or silencing,
the production of the enzyme (polygalacturonase) responsible
for the breakdown of pectin.[1] Though it turned out to be a commercial failure, Flavr Savr became a biocultural icon in the
debate over food security. Indian activist Suman Sahai, founder
of Gene Campaign, an organization dedicated to "the conservation of genetic resources and indigenous knowledge," invoked
the properties and developmental promises of the tomato.[2] In

"The Bogus Debate of Bioethics," Sahai makes a case for bio-
technologies in developing countries as an answer to food shortage, especially with regard to postharvest losses, while also
criticizing the export of first world bioethics to third world countries. She writes, "in India we must discuss the ethical aspects of
genetics or biotechnology rooted in our own philosophy and religion, reflecting our social and human needs, and resolving our
own dilemmas and problems in the way that is right for India."[3]
In a reply to Sahai, the ecofeminist philosopher and activist
Vandana Shiva stresses the importance of bioethics in developing countries and conversely rejects the biotech solution as serving the interests of global agribusiness and not people who need
the food to survive. Instead of bioengineered food importation,
she advocates locally grown products. As she put it, "the alternative is to reduce 'food miles.'"[4] Throughout this debate, the display
of a transgenic tomato that defies decay indexes competing geog-
raphies of self-sufficiency situated at the intersection of competing systems of obligation and translation. On the one hand is a
cultural obligation to rootedness powered by biotechnology and
reacting against bioethics. On the other is an environmental
obligation to rootedness powered by bioethics and reacting against
a biotech solution. At this point in the conversation, while waiting
for a third line of arguments and for a dialectic resolution, the
genetically modified tomato has morphed into something that
belongs to the realm of visual culture.

As a longer-lasting version of its more perishable counter-
parts, Flavr Savr is more image than tomato. One could even
say that the engineered fruit achieved in the context of a culture
of perishability the visual status of a still life. For most readers,
the term *still life* will conjure associations with Dutch painting,
small formats, a rather stern subject—arranged on a table, a

basket of ripen fruits on the brink of bletting—or with ostenta-
tious banquet pieces (*Pronkstilleven*), a more extravagant display
of luxurious delicacies, but probably not with genetically modified organisms. It seems almost perverse and certainly anachro-
nistic, and above all a paradox: nothing looks more like a tomato
than a genetically modified tomato. In fact, Flavr Savr will look
like a tomato longer. The anachronism—the troubled relation to
the time that passes—is not where we expect it.

In an attempt to break visual and historical molds (think of
all the fragile glassware items in Pieter Claesz or Willem Kalf
still life paintings), I propose to explore what it means to make
room for a biotechnological product in the history of representation. After all, the Flavr Savr tomato is an image of the future
of food and food security—albeit a deceiving one from the perspective of those who have lost their bet after having invested in
its promises. After all, biotechnology is already a major player
in the future history of mimesis.[5] To bring GMOs in the picture framed by still life is to think the future pictured by
biotechnology—the future of food security—according to a
different timeline and conceptual frame than the one generated by risk assessment or by a postmodern history of agriculture claiming that "biotechnology [has] been around almost
since the beginning of time."[6] It is to recognize that this picture
of a past always already opened on a projected future is internal
to a given structure. Continuity in time looks a lot different in
the creaturely meantime "of eating and drinking and domestic
life, with which still life is concerned."[7] For art historian Norman Bryson, whose analyses I follow here, still life is more than
a category used by art criticism, it is also "a response to . . . the
most entropic level of material existence" (13), and as such, by
virtue of a representational engagement with perishability, still

life can be understood as a venture in the biotechnological history of mimesis. Commenting on the lavish *Bouquet in a Niche* by Ambrosius Bosschaert (1573–1621), Bryson writes:

> All the flowers in the Boschaert exist at precisely the same moment in their life-cycle, when their bloom becomes perfect. The simultaneous perfection of so many flowers from different seasons banishes the dimension of time and breaks the bond between man and the cycles of nature. Which is exactly the point: what is being explored is the power of technique (first of horticulture, then of painting) to outstrip the limitation of the natural world (105).

Not only, still life painting instructs a present to linger over represented objects in a way we would never do over the actual objects, but it interrupts the continuity of a world of objects, appetites, and gestures directed at them. Still life painting defines the distance it puts between the pastoral background and a domain of technical (and commercial) achievement in terms of a representational (and imperial) effort.[8] The arrangement, floral or otherwise (shells, coral branches, and ornate vessels more commonly), abstracts the value of things that can be owned and consumed into things that can be painted, fixated, and framed. Or, in the words of Bryson, still life returns abundance and surplus to a rule: whether it is in the form of a taxonomic effect (a quasi-botanical arrangement of flowers), a virtuoso rendition of complex textures and intricate details, or an unassuming celebration of frugality and ordinariness.[9]

Let's consider by comparison an actual picture of genetically modified tomatoes. The source is an illustration from a scientific article comparing the decaying process, over ten, twenty, and forty-five days, of three types of tomatoes: a wild type (or

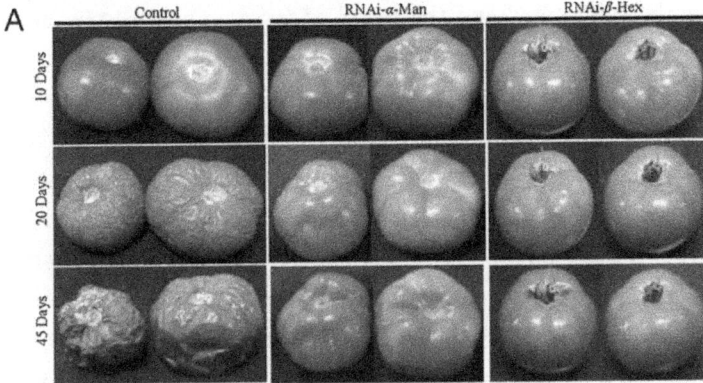

FIGURE 2.1 Vijaykumar S. Meli, Sumit Ghosh, T. N. Prabha, Niranjan Chakraborty, Subhra Chakraborty, and Asis Datta, "Enhancement of Fruit Shelf Life by Suppressing N-glycan Processing Enzymes," *PNAS* 107, no. 6 (February 9, 2010): 2413–18, figure 3A (2416). Reproduced with permission.

control group) and two transgenic specimens. The fruits are arranged in such a way as to frame a differential and compose a rather baroque tableau of transience. Here the image of the tomato as a *curiosité* to be placed in a biotech *Wunderkammer* next to other transgenic marvels (AquAdvantage™ salmon or Artcic™ apples) is unremarkable. A transgenic tomato is already an image of another tomato. Only time will tell the difference, qualify the image, prove a point, and, from an anti-GMO perspective, label a dangerously undistinguishable specimen. Therefore, what matters to the composition is the image of a process and a representation of time.

This process is not entirely foreign to Bosschaert's bouquet. A closer look will reveal the presence of insects: symbolic reminders of the mastered distance between the flowers in their respective landscapes and the flowers in the vase, between the

world represented and the world of representation. In an allegorical register, the fly is an image of transience and the butterfly a forbearing sign of resurrection. Likewise, a rapacious fauna
creeps in Balthasar van der Ast (1593–1657) and Jan van Huy-
sum's compositions (1682–1749). Insects at various development
stages (caterpillars, cocoons, and butterflies), spiders, amphibi-
ans, reptiles, and birds snack on precious flowers and pricey cit-
ruses. On occasion, they even prey on each other.[10] Decay lurks
everywhere in dots of blemish covering fruits and foliage. In
what Bryson refers to as the "still life of disorder," entropy also
takes its share with the messy spectacles of overturned vases
and of vessels on the verge of falling.[11] In still life painting,
decay and more generally entropy are the objects of a recoding
at the level of representation. Their manifestations are everywhere and yet put at a distance, creating, here too, a labeling
issue in art criticism that opposes panallegorist to antiallegorist
views. Bryson explains: panallegorists insist "on an absence of
naturalism and a saturation of the image by lexical codes," while
antiallegorists argue "for the absence of coding and an understanding of Dutch painting as a semantically neutral art of
description."[12] In the middle lies the vanitas effect.

I use this expression to underscore its dual nature as object
and project: vanitas as an assemblage of recognizable motifs
(skull, candles, broken glass, wisps of smoke, rotting fruits,
rapacious insects) and vanitas as illustration of a metaphysical
principle: "Vanity of vanities, all is vanity" (Ecclesiastes 12:8). A
vanitas painting is, to use the vocabulary of the theory of value
developed by Adam Smith, a "vendible commodity, which lasts
for some time at least after that labour is past." That is, as an
object, vanitas painting performs the recordability that distinguishes productive labor ("which adds to the value of the subject
upon which it is bestowed") from unproductive labor ("which

has no such effect"). At the level of the principle it illustrates, the vanitas effect undermines the time of accumulation and demotes principles of recordability. If everything—including the vanitas painting itself—is vanity, then the meditative services rendered by a vanitas ought to "perish in the very instant of their performance."[13] If properly executed, the trace it leaves behind ought to be discarded, self-destroying the image in the process, consumed but without appetite.

In the same way that for the art historian Ernst Gombrich the vanitas effect appears built-in in every painted still life "for those who want to look for it," metabolic and bacterial triggers of decay are built-in in the food on display in Dutch still life, yet invisibly and nonallegorically.[14] Unlike in the realm of still life representation, gene silencing does not remove the tomato from organic time. Bioengineered or not, the tomato will eventually rot. But because the gap between the signified ("the activity of polygalacturonase causes the breakdown of pectin" or "everything rots, everything is vanity") and the signifier (gene silencing, on the one hand, and the visual performance of still life, on the other) has been built into the Flavr Savr's flesh as its principle, the internal contradiction between sign and matter is lived, or rather metabolized, so that it "leav[es] an imprint on a dynamic bodily process."[15] The vanitas effect takes advantage of the coming of the world to a still as it makes apparent the gap between sign and matter, between the "what" and the "how" of the picture. But the image itself does not solve or resolve the divide between the thing and the sign. It exists through the divide itself both as allegorical form and as a pictorial telltale of a breach in time watching over the instability between meaning and being.[16]

No matter how you handle them, the oysters in Jan Davidsz de Heem's *Still Life with Oysters and Grapes* will not feed you. Although as signs it has no nutritional value, represented food can still play a metabolic function as a mechanism of control in the same way that for Simon Schama Dutch still life painting opened "a dialogue between [a] newly affluent society and its material possession."[17] Charles Sterling implicitly acknowledges the metabolic function of still life painting in the conclusion of his monograph on the subject, but to a different end:

> Today we are surrounded by unparalleled quantities of machine-made objects. We live, as it were, redoubled by objects which prolong the effective action of our limbs and extend our field of vision. But though we create them ourselves, their substance and forms are alien to us and take us by surprise. The mechanical processes of their manufacture remain enigmatic for us and leave us indifferent. They have *not yet* found a place in our inner life.[18]

Prospects are not good for still life painting. Biotech displays will not do: it is not enough to introduce new objects, such as a genetically modified tomato, into the picture. The picture has to digest them so that they can find "a place in our inner life." What such a process of internalization entails, especially in the context of gene transfer technology, is far from clear. But for Sterling it registers time. It designates a border between the future (of the still life painting genre itself and that of its reflexive-normative properties in a world of prosthetic affordances) and a meantime—the "not yet"—of representation. A sense of continuity in time—a time in which objects still belong to us and where prosthetics complete us ("prolong the effective action of our limbs and extend our field of vision") rather than dilute us—is up for grabs. And yet for Sterling the future of the still life seems to be out of reach. As Bryson also observes in Willem Kalf's painting: "because it has lost its bond with the actions of the body, matter [in still life painting] is permanently out of

place. Its spatial co-ordinates are those of acquisition, naviga-
tion, finance: theoretical axes with which the body can never
intersect. . . . *Still Life with Metalware* [1644] shows the true
resting place of the object fully commodified and abstracted:
the table is like a bank-vault—or a graveyard."[19] If one accepts
as premise that the history of still life painting is not over but in
fact does inform the very perishability of Flavr Savr, then it
stands to reason that the list of spatial coordinates that defines
modalities of being within or out of reach is not closed. It goes
on with gene silencing and, in the broader context of the governance of loss, storage, and decay, should include delayed germination, cryopreservation, and protocols of termination.

The future of still life is in the "not yet" of the Svalbard Seed
Vault project described in the previous chapter: "The Seeds [sic]
in the Seed Vault shall only be accessed when the original seed
collections have been lost for any reason." It is an image from
within the postapocalyptical crypt of history. It offers a vision of
what Svalbard's seed chambers would look like once the doors
are closed, in between seed deliveries, or after the doomsday
event—whatever it turns out to be. It is less an actual image of
what will come or what is inside than a reminder of the insurance policy that is the vault to come into effect. In the meantime, the role played by the vault in "reducing hunger and
poverty in developing countries" lies in the creation of a reser-
vation status and in the number of species preserved.[20] What
is transmitted to the hungry future are not seeds, but genetic
diversity. Genetic diversity may amount to food at the end, but
again, in the meantime, the vault's existence and historical
significance feeds on a metabolic paradox: although seed banks
stabilize biodiversity and secure food futures, for many, in
the context of the American biopolitics of morbid obesity
and impoverished agency, food is less about the future as the

condition of living on than it is about the intensified codependence between pleasure and finitude.[21] Some creatures, however, seem to be thriving in that ecology of distress. Homer Simpson's survival in the midst of a heavily polluted environment, and his exceptional endurance as a cartoon character feeding on transfats and high fructose corn syrup, defies all epidemiological odds. It is almost as if he were fictional in metabolic terms rather than on the basis of his avowed animated existence. It is against this background, constituted by the interiorization of disaster defining Svalbard as project, that it is possible to envision the future of still life and something like a still life with genetically modified organisms.

From still life to the engineered modification of shelf life, there is no solution of continuity. To call on the future of early modern Dutch still life painting in the present of food security is a way to unsettle the terms of the Sahai/Shiva debate, first, in the sense that to treat a GMO tomato as an image is neither to say yes or no to a biotech solution. It is to stick with the making and unmaking of a sense of continuity in time. It is to give a new sense of depth to the future pictured and debated through bioengineering and its discontents. Second, it serves as a reminder that GMOs do not feed the future. They are representations, and representations will not feed anybody (even in the case of a picture made with edible pigments, what ends up being consumed is not the representation itself but its materiality). The future, if it is a representation, is not something that needs to be fed, or only metaphorically. Conversely, if the future is hungry, it is only by representation—which leads us to the last point. What we have here, faced with the visual dimension of a biotech product, is the requalification of the history of representation as a strain of discourse—one among many, and certainly not the most policy relevant—on the governance of decay.

The future pictured by GMOs is only a future if a present manages to envision it in relation to a past. In this chapter the past is not the one defined by the convergence between histories of plant breeding and genetics.[22] It is a cultural memory of transience that draws a dissenting line between the history of "creaturality" in art and the future of perishability in the biotech present.

3

STORE AND TELL

The principle which prompts to save is the desire of bettering our condition, a desire which, though generally calm and dispassionate, comes with us from the womb, and never leaves us till we go into the grave.

—Adam Smith, "Of the Accumulation of Capital, or of Productive and Unproductive Labour," *The Wealth of Nations* (II.3)

Adam Smith's views on frugality are not the object of this chapter, only a point of entry into the difficulty in distinguishing between storage and burial. The point of entry is not the claim to a form of anthropological truth itself—the quiet but efficient drive to store we inherit at birth only leaves us with death in Smith's narrative—but its sententiousness. For Marx, taking direct aim at *The Wealth of Nations*, this statement of principle is part of what he describes as the legend of primitive accumulation that sets the "frugal elite" apart form the "lazy rascals."[1] It is, he pursues in the same paragraph, but a story for kids (*kinderei*), that is, to serve something like an education of desire that the very sententiousness of the lesson denies in principle (the desire to save is natural). In

Smith's statement, the distinction between storage and burial is clear and in a sense reassuring. There is a beginning, there is an end, there is continuity between the two, and there is a foundational quality to it, no matter how childish. Other views alert us to more nightmarish and uncanny versions of the story: after all, what if the grave was but another kind of matrix engendering unclassified forms of life and animating unrecognizable desires? What if accumulated stuff took on a life of its own once deposited in the grave of consumerism? Landfills teem with dangerous bacteria and toxic compound.[2] Stored stuff has a life that is not deemed sustainable in present circumstances, sometimes because of the limited space available in certain highly competitive living conditions.[3] Or sometimes because it is radioactive.

What fables of perishability teach us is that storage and burial are two variables in a continuum of operations, not fixed attributes. The distinction seems to depend less on the structure of the container itself than on what it actually contains: stored life and buried death, vessels full of wheat or vessels filled with bones.[4] Though their respective modalities might be the same, burial and storage assumes rather different metabolic functions. A burial is thought of as unidirectional—what is buried once is forever buried. This is why the spectacle of forensic TV shows digging up bodies for the potential clues they will offer and walking corpses in zombie movies are so disconcerting; they steal from coffins and burial grounds their sense as a final resting place and turn them into a body storage device. Through its commitment to the remaking of life and death, biotech radically transforms the terms of the opposition between storage and burial. The ending of Julia Leigh's novella *The Hunter* (1999) is quite telling in this regard.[5] Throughout the story, the reportedly extinct thylacine (or Tasmanian tiger) exists as a collection of traces left by a hypothetical last specimen. The coveted animal materializes itself in the last few pages and just long enough to be shot dead and extinguished a second time. The last thylacine receives a proper burial, but only after a thorough harvesting of it genetic material (blood, ovaries, and uterus). Here the burial is a way to ensure exclusivity over stored tissues and secure their potential biotech value.

Media history tells a different story. Alexander Klose suggests that in the context of a constant worldwide circulation of containers, the difference between storage and burial recedes in the opposition between chest and box. He writes in *The Container Principle*: "In the chest, the history of the objects that it contained (and therefore the history of its owner) has been saved. In the box, all histories are reconfigured with each load."[6] There seems to be no room for eternal rest in this geopolitical state of affairs, or perhaps only with the collapse of international trade. Klose's distinction between chest and box adds up to the distinction between *pithos* and barrels. Unlike movable barrels, *pithoi* are large vessels buried halfway in the ground and serve as storage unit for wheat, wine, or bones. He insists that, in the Greek mythology, Pandora does not open a box (*pyxis*) but a *pithos* containing all the *kaka* of men, "labor, the fatigue from work, the suffering associated with them, and last, disease . . .—everything that saps the strength of mortals, ruins their vitality, and makes them age and die."[7] French Hellenist Jean-Pierre Vernant confirmed this idea in his reading of Hesiod while insisting on other properties of the *pithos* as unit of time.[8] Sealed and full of "wheat" (*bios* in Hesiod's text), *pithoi* contain the future of domestic continuity. Opened, they signal the present of consumption. Pandora's incident itself marks a difference in time between the golden age (before the invention of the first woman, Pandora) and the subsequent ages of decline. By extrapolation, the difference between storage and burial locates problems that have to do with the disciplining of a certain experience of continuity between past, present, and future. For instance, if historiography, according to Michel de Certeau, buries the dead "as a way to establishing a place for the living," by contrast, sustainability, in Leerom Medovoi's words, "seeks to gauge the kind and amount of life that must not be killed now so that the process of surplus value extraction can continue indefinitely into the future."[9] This particular metabolic ambivalence between burial and storage is particularly salient in Charles Perrault's version of the Bluebeard folktale (1697).

As the story goes, Bluebeard's murdered wives are not buried, but ghoulishly stored in a chamber (*le petit cabinet*) somewhere in his castle, their corpses hanging from the walls and their blood pooling on the floor to form a reflection pool. This scene is not self-contained. Like the rest of his wealth encased in coffers, strongboxes, and dressers, it demands to be seen. It is as if the very vision of the inside of the chamber were making storage room for more victims. By structural contrast, the primal murder and its logic, the how-it-was-done-in-the-first-place, is buried and unaccounted for in the text.[10] The bloody chamber has been visited an unaccountable number of times. Its blood pool captured and continues to capture: the same scene and the same search for what remains to be seen. This is the story of a leak, not an accidental leak to be solved by technical means, of a leak bearing witness to an architectural and technical failure to contain, but the story of an organized leakage instrumental to an unfolding cycle of violence. We know this because we are watching as it is to happen again.

Bluebeard entrusts his new wife with the keys to the chambers and coffers in which he accumulates and displays his fabulous wealth. But as he leaves the castle he tells her that one of them—the key to the small cabinet—should not be used. She rushes to open the forbidden chamber, sees the dead ex-wives reflected on the blood-lacquered floor, and drops the key in the blood. The stained key cannot be cleaned and betrays her when she hands it over to her husband upon his untimely return. The verdict is without appeal. She is fated to join her predecessors in the chamber to feed its storing logics and to become, in death, the next Mrs. Bluebeard. However, with the help of her siblings, she manages to break Bluebeard's cycle of violence and restore the possibility of transmission in a system where a form of homicidal-bound accumulation had prevailed.[11] Because Bluebeard has no heirs—a direct result of the lethal logic of his alliance—she inherits his wealth, which she uses to endow her siblings and later marry herself to "quite a gentleman" (*un fort honnête homme*). The last sentence of the tale informs us that her new husband "ma[kes] her forget the ill time she had passed with Bluebeard" and buries her story in oblivion, as if, in order to restore transmissibility, the story had to annul itself at the end. Quite remarkably, the ending of her story as Mrs. Bluebeard mirrors its beginnings, marked by the erasure of the previous wives stories ("No one knew what happened to these women") and by the disappearance of the signs of warning ("all went so well that the younger sister started to believe that the lord of the manor did no longer have so blue a beard and that he was quite a gentleman [*c'était un fort honnête homme*]").[12] In the concluding sentence of the tale, the ordeal of the survivor has almost the same status as the first unaccountable murder in Bluebeard's cycle of violence.

As one storage operation closes, another opens. Two appendixes to the tale indicate that a moral load has been stored in the story. Like Bluebeard himself with the help of the forbidden fairy key, the two *moralitez* stage a failure to comply. The first morality reads, "La curiosité malgré tous ses attraits, / Couste souvent bien des regrets; / On en voit tous les jours mille exemples paroistre" [Curiosity, despite all its appeal / Often costs many regrets / One sees every day thousands examples of it appear].[13] The second morality reads, "On voit bien-tost que cette histoire / Est un conte du temps passé; / Il n'est plus d'époux si terrible" (82) [One sees very soon that this story / Is a tale of times past / There is no more husband so terrible"]. The continuity that the present ("On en voit tous les jours") presupposes in the first morality is undermined by the very historicity that the second moral appendage claims: "There is no more husband so terrible . . ." The story therefore exists in the past by virtue of a once-upon-a-time, but in a relation of irrelevance with a current state of affairs.[14] Because the *voir* (as vision, sight, and recognition) recedes into the appended present (*il n'est plus*), the future will not unlock anything. It seals the cautionary tale onto itself. There is nothing left to see. In "The Blue Beard," everything is already visible. Each door has its key. Upstairs, the mirrors are large enough to serve as reflexive caskets ("mirrors in which one can see oneself from head to toe" ["des miroirs où l'on se voyait depuis les pieds jusqu'à la teste"]). In Bluebeard's castle the desire to see is stronger than the promise of destruction: "There is nothing that you should not expect from my wrath" ("il n'y a rien que vous ne deviez attendre de ma colere"). The problem is not to sneak in to see but to leave the castle

alive, to put the past behind, to survive the scene that was concealed and yet offered to disclosure. The same problem resur-
faced under a different form in the middle of the twentieth
century under different conditions that have to do with the governance of decay.

In 1980, the U.S. Department of Energy (DOE) convened the
Human Interference Task Force (HITF) to "investigate the
problems connected with the postclosure, final marking of a
filled nuclear waste repository. The task of the HITF is to devise
a method of warning future generations not to mine or drill at
that site unless they are aware of the consequences of their
actions."[15] Located near Carlsbad, New Mexico, the Waste Isolation Pilot Plant (WIPP) site is destined to accommodate
175,000 cubic meters of transuranic waste stored 2,150 feet
underground in rooms excavated in salt sediments that will play
the role of a sealing agent—sealing off the relation of a past
made of programs of military defense leftovers to the historical
present.[16] The seal also warrants a form of continuity in time
that potential leaks in the WIPP structure could affect. As
Peter van Wyck pointed out, the repository cannot be so well
concealed that its location falls into oblivion. In other words,
the burial ground—"a burial without mourning"—must remain
a storage unit.[17] A symbolic leak in the form of "signs of danger"
is required to prevent an unintended toxic spillage.

Georgius Agricola expressed similar concerns in the sixth
book of his monumental treaty on the arts of mining and met-
allurgy published in 1556 in Latin (and made available in English
by none other than President Herbert Hoover and his wife Lou
Henry):

Shafts and tunnels should not be re-opened unless we are quite certain of the reasons why the miners have deserted them, because we ought not to believe that our ancestors were so indolent and spiritless as to desert mines which could have been carried on with profit. Indeed, in our own days, not a few miners, persuaded by old women's tales, have re-opened deserted shafts and lost their time and trouble. Therefore, to prevent future generations from being led to act in such a way, it is advisable to set down in writing the reason why the digging of each shaft or tunnel has been abandoned, just as it is agreed was once done at Freiberg, when the shafts were deserted on account of the great inrush of water.[18]

Written instructions are not enough or at least not as straightforward a solution on the long run. In his report to the Office of Nuclear Waste Isolation, Thomas Sebeok, a HITF member and at the time a professor of semiotics at Indiana University, came to the conclusion that "no fail-safe method of communication can be envisaged 10,000 years ahead."[19] He introduced the idea of a redundant system of relays that should be periodically updated so as to counter the effect of negative entropy. The report recommends that information regarding the transuranic waste burial

be launched and artificially passed on into the short-term and long-term future with the supplementary aid of folkloristic devices, in particular a combination of an artificially created and nurtured ritual-and-legend. . . . The legend-as-ritual, as now envisaged, would be tantamount to laying a "false-trail," meaning that the uninitiated will be steered away from the hazardous site for reasons other than the scientific knowledge of the possibility of radiation and its implications; essentially, the reason would be accumulated superstition to shun a certain area permanently (24).

In a sense, Sebeok reverts to a counterversion of Agricola's "old women's tales." What led him to this recommendation was none other than the memory of Pandora's jar, cask, or vase (but not box, Sebeok insists), a cultural memory that, in effect, spans millennia. However, the report also acknowledges that

> the best mechanism for embarking upon a novel tradition, along the lines suggested, is at present unclear. Folklore specialists consulted have advised that they know of no precedent, nor could they think of a parallel situation, except the well-known, but ineffectual, curses associated with the burial sites (viz., pyramids) of some Egyptian Pharaohs, e.g., of the 18th dynasty, which did not deter greedy grave-robbers from digging for "hidden treasure" (24).

Bluebeard knows this all too well. In the wake of a toxic transuranic burial, Perrault's text—another example of "folkloristic device"—portends this failure to warn and be warned. "Old women's tales" can ward off intruders or push inquisitive minds to reopen deadly mines. There is no telling.

Sebeok's model is a striking example of what a security system negotiated in diffused participatory terms, rather than in expert-reliant techno-architectural terms, would look like. The former executive director of the Global Crop Diversity Trust, Cary Fowler, observes a comparable phenomenon with the Svalbard Seed Vault:

> I remember when we were constructing the facility and I was talking to the local Governor in Svalbard who's responsible for

security on the island, and he said to me "Cary, if anyone so much as writes graffiti on this thing we'll know who it is." After all it's just a small village there, and really what's neat is that the villagers are remarkably proud and protective of the vault. They know it's there, and they're proud of it and that gives us an extra security blanket out there because the locals see everything that's going on—walking around up there I've had any number of people stop me and say "we're protecting that vault of yours."[20]

The resurrection imaginary embodied by the seed banks and the transmission logic envisioned for the WIPP by Sebeok seem to have unexpectedly fused in a passage from Walter Benjamin's essay on the end of storytelling and the destruction of experience: "Herodotus offers no explanations. His report is the driest. That is why this story from ancient Egypt [Psammenitus's story] is still capable after thousands of years of arousing astonishment and thoughtfulness. It resembles the seeds of grain which have lain for centuries in the chambers of the pyramids shut up airtight and have retained their germinative power to this day."[21] Much has been said about "The Storyteller," and it is not my intention here to add to the commentaries. I just want to point out that, if the age of information has replaced the art of storytelling, as Benjamin argues, Sebeok's search for folkloristic precedence in matters of transhistorical security apparatus and control mechanisms offers a new version of this scenario, retrofitting a storytelling tradition to solve an informational problem posed by a transmillennial nuclear deposit.

The DOE has since opted for Passive Institutional Controls (PICs), which include surface and subsurface warning markers, a monumental berm, and multiple archival repositories, both on- and offsite.[22] But regardless of its current status, Sebeok's proposal survives as a striking experiment in cultural sustainability challenging both the engineering response to a communications problem and the institution of literary studies as a form of remembering. It challenged me to reopen "The Blue Beard" to reconsider its contradictions (the story belongs to our time/the story belongs to the past). The text survives in a contaminated present as the paradox of a cultural memory of erasure. In the context of what Myra Hird has called "terminal capitalism," Perrault's tale of secrecy and power becomes a tragedy of cohabitation. That is, in conditions "whereby our only solution for dealing with toxic and contaminated material that our relentless consumption and planetary depletion generate[,] is by producing permanently temporary waste deposits for imagined futures to solve," to run the risky business of keeping corpses in the castle takes on a different meaning.[23] It is no longer part of an elaborate power trip.[24] Making the corpses disappear is simply not an option.

Sebeok's archived proposal responds in an anticipatory mode to literary critic Hans Ulrich Gumbrecht's unforgiving statement in an essay provocatively titled "Shall We Continue to Write Histories of Literature?": "After all, whether we are ready to admit it in public or not, literary critics know all too well that humankind would easily survive without literary criticism—and most likely even without the humanities at large."[25] In the deep future of nuclear waste management envisioned by Sebeok's *Communication Measures to Bridge Ten Millennia*, survival might well be a matter of renewed critical engagement with notions of literary tradition and philological expertise. Or maybe it is that survivalism is, after all, a form of literary culture.

In Richard Mitchell's ethnography of militias and paramilitary groups in North America, survivalism has defined its own mode of relating to printed words and their circulation in flyers, magazines, recorded tapes, diaries, and fiction, and modes of acting out responses to perceived threat. Survivalist literacies answer the questions "Who shall create? Who shall have a hand in crafting culture for the twenty-first century?"[26] Survivalism is thus not merely what's left when dire necessity has removed the ornamental, the surplus, the foam of the unnecessary, the froth of the literary—whatever the name and the categories epics of futurity reserve for irrelevance and unfitness to endure postapocalyptical times. Rather, it somehow takes up Bill Readings's demand in *The University in Ruins* urging students of posthistorical literary culture to find other ways to understand how what we say about literary texts and culture participates in culture, once "the decline of the nation-state as the primary instance of capitalism's self-reproduction has effectively voided the social mission of the modern university."[27] Survivalist literacies anticipate the end of a politico-cultural project insofar as they commemorate a time in which discipline and piety could be found in a particular arrangement of narratives and bodies that live to fight and tell. In this respect, survivalism constitutes, in the history of the modern political imagination, a form of social contract "centered on the continuing task of constructing 'what if' scenarios in which survival preparations will be at once necessary and sufficient."[28] These storylines turned lifelines anchor forms of agency (hoarding, hiding, budgeting) and values (autonomy, family, craftsmanship) in a geography of safety (mountains, woods, camps) and threats (invasion, cyberattack, recession, fuel depletion, racial secession) replacing a geopolitical order soon to be overthrown in an anticipatory tomorrow (what once was Colorado and is now known as . . .). What-if scenarios inhabit a present comprised of autonomous fighting cells ready to move to their bunker or already on the move as they rehearse for an impending disastrous event: building shelters, hoarding weapons, and stockpiling bottled water, canned food, and spare batteries.

Survivalists tell stories blurring the distinction between words and deeds, life and fiction (hence the disturbing dimension of some of Mitchell's first-hand accounts). They tell one story in particular: the last one, the one that counts, and recounts from the end, how a community became radically minoritarian because it will have been the last one—or one of the last ones—standing, entrusted with the literary task par excellence to serve as social and cultural memory. Through their obsession with apocalyptic fables, camouflage, proofed shelter, sealed arks, and concealed architectures, survivalists interiorize an ending to keep it from ending. In the final section of *Dancing at Armageddon,* "North," Mitchell offers a last report on his expeditions into the world of survivalism. This project brings him to Canada where he visits a gigantic private antinuclear shelter in Ontario, twice. The first visit ends with a striking vision: "We emerge into the sunset, time travelers back from a possible future. . . . Sand's ark to the postapocalypse had already safely carried precious cargo to the future. The whole complex honeycombing the hillside, unit by unit welded together, entrance to entrance, back to front, right and left, was made from forty-two wheelless, cement-encased, faded yellow school busses, ready for one last hopeful trip toward a better tomorrow."[29] Years later, Mitchell comes back for a second visit. The vandalized ark is rusting away, still waiting for the future that left it behind as the emptied shell of its trashed survivalist gest. The illustrations gracing the cold war efforts to normalize and give a sense of bearablity to life in a postapocalyptic world share the same fate.

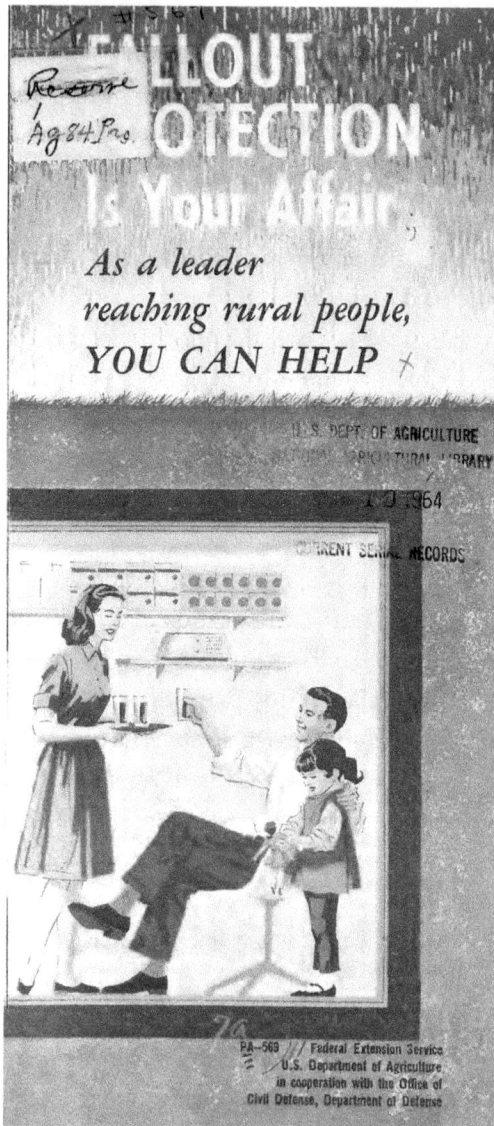

FIGURE 3.1 *Fallout protection is your affair.* Washington, DC: Federal
Extension Service, U.S. Dept. of Agriculture, 1963.

Federal Civil Defense Administration (FCDA) publications and fallout shelter handbooks offer views of livable mid-century interiors populated by white middle-class nuclear families. In these fallout shelter views, the eye peels away layers of contaminated earth to reveal tableaux of blissful domesticity and oblivion. It positions a not-yet in a past, which came to imagine its future underground, tucked in a place where, Joseph Masco noticed, "there is no sign here of the reality of nuclear war, of the scorched and barren radioactive landscape, of the extreme trauma of life in a postnuclear environment."[30] These illustrations are archival in the sense that they image a continuity in time in relation to an imagined future that nevertheless survives as an image from the past. It is as if life after a potential nuclear blast demanded to be seen, not to be less unbelievable but rather for the thought of survival to appear less unbearable. The scene from within the underground structure is a site from which to remember and mourn the scorched surface to be viewed as through a window. But the only window left in the aftermath of a nuclear holocaust is the one these images opened onto the shelter itself.[31] In the cold war fantasy of life underground, as in Leon Battista Alberti's Renaissance treatise on painting, the window is a metaphorical aperture.[32] The key passage in his *De Pictura* reads: "on the surface on which I am going to paint, I draw a rectangle of whatever size I want, which I regard as an open window [*aperta finestra*] through which the subject to be painted [*historia*] is seen."[33] The illustrations from the FCDA are themselves windows opened onto a world in which they have been replaced by collapsable tunnels and blast doors. If there are openings left to see through in this world, if there is a sense of perspective left in it, it is as a breach that is likely to be lethal.

Some of the most sophisticated underground interiors—the shelter design by Dorothy H. Paul for instance[34]—contain books

FIGURE 3.2 Georgius Agricola, *De Re Metallica* (Basel, 1556), 73. Courtesy
of the University of California Libraries.

FIGURE 3.3 Georg Agricola, *De Re Metallica* (Basel, 1556), 160.
Courtesy of the University of California Libraries.

and pictures. Who knows what volumes sit on the shelves?
Nothing will prevent me from imagining that Agricola's treatise
next to Perrault's *Mother Goose Tales* did survive an extinguish-
ing nuclear blast. The woodcuts illustrating the *De Re Metallica*
(1556) have already opened a window onto the subsurface. In
these images the landscape bleeds into the excavated ground to
offer views of a laboring humanity whose tunnels have turned
depth into a storied volume. In book 6, Agricola goes on to
describe ingenious ventilating mechanisms devised to make
mines breathable but hardly livable. Toxicity lies underneath in

many forms: lethal vapors, corrosive minerals, and also demonic forces: "In some of our mines, however, though in very few, there are other pernicious pests. These are demons of ferocious aspect, about which I have spoken in my book *De Animantibus Subterraneis*. Demons of this kind are expelled and put to flight by prayer and fasting. Some of these evils, as well as certain other things, are the reason why pits are occasionally abandoned."[35] The folkloristic is never far, even though it does not seem to have left traces in the illustrations. It lingers on, not as a geological reality, but as a repelling mechanism Sebeok is not afraid to summon in the midst of the cold war.

One of the illustrations of *De Re Metallica* resurfaces in Danish director Michael Madsen's 2003 documentary on the Onkalo, Finland, subterranean nuclear waste repository. The image is cited as an example of a preindustrial excavation achievement. The engraving is the image of a technological past that, tens of millennia into the future, could very well break Onkalo's shield and release its toxicity. For the engineering team in charge of securing the Finnish repository, the effect of continuity between surface and subsurface—between *historia* ("the subject to be painted") and geology—depicted in the *De Re Metallica* becomes at some point in the distant future a potential form of continuity in time.

The association between toxic waste disposal and management, ecological disaster, and the terminal future of the humanities congeals in a conservation conundrum. Discussing the preservation of "the task of thinking" within instructional structure in the mid-nineties, Readings writes:

> It is not a question of coming to terms with the market, establishing a ratio of marginal utility that will provide a sanctuary—such

a policy will only produce the persistant shrinking of that sanctuary, as in the case of old-growth timber in the United States.
How many philosophers, or redwoods, are required for purposes
of museification? If both the grand project of research and the
minimal argument of species preservation are likely to prove
unsuccessful, it seems to me necessary that our argument for certain practices of thought and pedagogy must measure up to the
situation and accept that the existing disciplinary model of the
humanities is on the road to extinction.[36]

In other words, if research will not save the humanities, as
Readings makes clear earlier in his article, then what "will
repay the costs in time and capital expended" by humanistic
disciplines in the name of liberal education?[37] Will generous
gifts, estate donations, and private endowments save, for now, a
culture of humanistic inquiry in research universities?[38] Salvage
rhetorics miss Readings's exacting point: "Thought is nonproductive labor, and hence does not show up as such on balance
sheets except as waste."[39] The question is thus neither "Do the
humanities have a future?" nor "What claim do the humanities,
or scholarship generally, have on increasingly limited resources?"[40]
Rather, what are the forms and practices under which, waste,
decay, and nonproductive labor register on the balance sheets of
terminal capitalism? Sebeok's proposal for a culture of toxic
guardianship is one of these forms. It is now but an archived
remain. Two other models seem alive and well today.

　　Hoarding on TV exists as a familiar plotline in the current
drama of sustainability where an overwhelming form of non-
fungible accumulation strangles social relationships and threat-
ens to tear a home apart. Stuff took on a life of its own in what
used to be the living room. It turned the kitchen into an
unlivable closet space. Filled with junk, the bedroom cannot
accommodate intimacy anymore. Backed-up corridors no longer

facilitate passage. Hoarders walk the tenuous line between storage and burial. For them, the difference between the two is not an abstraction. It is lived and defines a present. In the TLC reality show *Hoarding: Buried Alive*, a deus ex machina with a handful of garbage bags or a loading truck will restore the domestic space to a sense of normalcy. For a decluttering intervention to occur, however, one has to believe in the availability of a world outside, where the story of the traumatic event behind the hoarding habit can be uncovered and where stuff can be dumped indefinitely.[41]

In the *Larousse* French dictionary, "To glean is to gather after the harvest." In Agnès Varda's documentary film *The Gleaners and I* (2000), to glean is to intervene in the governance of perishability.[42] It is to puncture the life cycle of consumables organized by expiration dates, production quotas, and the calibration of potatoes. Gleaning transforms ownerless stuff into transmitted objects. From the fields where unmarketable crops are left to rot, to the street curbs where refuse and old appliances pile up, Varda gleans stories of gleaners whose salvaging gestures actively and creatively contest ways of letting things go that are embedded into wasteful modes of production and consumption. One of these gleaners, Bodan Litnianski, is a retired mason. Over the course of his adult life, Litnianski turned his garden into a sculptural maze made of collected debris. Among the cemented objects composing what he calls his "system," he encased an entire population of dolls in improvised niches. Weathered, beheaded, maimed, and variously defaced by the decomposition of their polyurethane flesh, his dolls are not forever young, forever dolls, forever toys. They exist in a new frightening life cycle. Litnianski died in 2005. Since the passing of his wife Emilie three years later, his fragile ecology of rotting dolls has been abandoned to untamed vegetal growths.

The creation of a protocol of compassionate disposal for frozen embryos mentioned in the preface belongs to the same register of practices that generates, reflexively or not, scripts and
scenarios to bear the cost of perishability, and so does the biography that I propose to examine in the following chapter:
Rebecca Skloot's *The Immortal Life of Henrietta Lacks* (2011).[43]
Tissue culture redraws the lines between waste and resource
by assigning value to the management of decay and other complex metabolic functions. This means that genetic information,
blood, scraps of organs and metastatic tumors can, in some circumstances, become valuable. This also means that, in a striking
reversal of the process of biographical capture, the making of
biovalue can generate biographical remains. And as such, in a
culture of perishability, the lived mortal life of Henrietta registers as a waste by-product of her immortality as HeLa.

4

THE MORTAL LIFE OF HELA

I would never recognize her other than in pieces.
—Roland Barthes, *Camera Lucida*

In telling the untold story of an African American woman, whose biography had been eclipsed by her encounter with the history of biomedicine, Rebecca Skloot's *The Immortal Life of* Henrietta Lacks had perhaps the unexpected effect of accentuating the scalar division between the human and the cellular. For HeLa, an in vitro organism, has a life of its own. It is the mutating cellular remnant of Henrietta Lacks's cervical cancer biopsied at Johns Hopkins Hospital in 1951, shortly before her death. HeLa cells are immortal in the sense that, unlike regular cells, they undergo cellular division indefinitely due to a certain level of telomerase activity that prevents the shortening of telomeres responsible for cellular death (or apoptosis). In *The Immortal Life of* Henrietta Lacks, cellular immortality is not simply a physiological oddity or marvel, it is a milieu of intersection charted by biomedicine but endured as biography. In this milieu, critical junctures occur as instances of the transformative relations that exist between what Arjun

Appadurai calls the "peculiar half-life of any piece of reliable new knowledge" and the fact that research projects are designed and carried out on the assumption that they shape bodies and impact lives.[1] More important in the context of this book, cellular immortality is a present with its own duration and punctuations. As a "case," that is, as a story told so as to sanction a certain range of clinical, legal, and bioethical interventions, the life—mortal and immortal—of Henrietta Lacks remains in limbo.[2] The case is all but closed. Its vitality cannot be dissociated from the presence of telomerase activity that warrants both endless cellular divisions and experiments in perishability.

If the transnational velocity of HeLa's trajectory is indifferent to the ordinariness of depression and survival in postplantation America, some, however, have to live with this indifference. On HeLa's side of the story, the fast-paced biopic of tissue culture and postgenomic medicine delivers futurity and experimental promises to the world. On Henrietta's side, the past lives on through narratives of exhaustion in a paradoxical present shaped by a legacy of dispossession and destitution: "while HeLa is regarded as the 'mother' of various medical advancements, Henrietta's motherhood yields no financial benefits for her heirs." Marlon Rachquel Moore continues: "This irony is not new, for Black women a complicated history of maternal and reproductive rights that further links HeLa's biofuture to this country's racial past."[3] For the Lacks's family, subjected to further medical tests but left without healthcare, cellular immortality is a by-product of death, social deprivation, and systemic violence; for them, the history of biotechnology remains attached to a sociopolitical order that, in Hortense Spillers's words, "*represents* for its African and indigenous peoples a scene of actual mutilation, dismemberment, and exile."[4] On Henrietta's side of the story, cellular immortality is a paradox of the "flesh," a term that Spillers refashioned in her seminal 1987 article. Her conceptual reappropriation does historical work while questioning a history predicated on the caesura between past and present, life and death, corporality and personhood. Flesh is constituted in the skinning alive of the captive body. There is flesh because capture and torture do not simply kill. They turn beings into things, persons into objects. They transmit objectification. In other words, there is flesh because "the person of African females and African males registered the wounding."[5] The story of this objectification is not over yet: in her pioneering work on the development of tissue culture, Hannah Landecker has shown how the personification the HeLa cells was underwritten by a process of racialization.[6] More recently, Alexander Weheliye has noted that: "If Henrietta Lacks's story and the ongoing narrative of the eternal life of the HeLa cells prove anything, it is that the hieroglyphics of the flesh subsists even in death, and that it has now been transposed from the outwardly detectable to the microscopic interior of the human."[7]

The flesh is an event that does and does not register in the history of the notion of person sketched out by Marcel Mauss in his 1938 Huxley Memorial Lecture. Yes, Mauss knows that in ancient Rome a slave is not a person: "*Servus non habet personam.* He has no 'personality' (*personnalité*). He does not own his body, nor has he ancestors, name, *cognomen*, or personal belongings."[8] But there is no explicit mention of American slavery in the rest of the lecture. I would suggest nevertheless that the concluding section is haunted by the thought of its defacing effects:

> Who knows even whether th[e] "category" [of person], which all of us here believe to be well founded, will always be recognized as such? It is formulated only for us, among us. Even its moral strength—the sacred character of the human "person"

(*personne*)—is questioned, not only throughout the Orient, which has not yet attained the level of our sciences, but even in the countries where this principle was discovered. We have great possessions to defend. With us the idea could disappear.

(22)

The specter of the flesh is made even more haunting in Mauss's text by the fact that it is already at work in the phrasing itself and its insistence on the sense of ownership ("We have great possessions to defend") that both warrants and guards access to the category of person. The flesh has already interrupted that history by anticipating its "sketchy" premises and its plasticity. Mauss explains in his introduction:

> What I intend to do is to provide you with a summary catalogue of the forms that the notion [of person] has assumed at various times and in various places, and to show you how it has ended up by taking on flesh and blood, substance and form, an anatomical structure [*arêtes*], right up to modern times, when at last it has become clear and precise in our civilizations (in our European ones, almost in our lifetime), but not yet in all of them. I can only rough out the beginnings of the sketch or the clay [*argile*] model. I am still far from having finished the whole block or carved the finished portrait.

(2)

The immortal life of Henrietta Lacks could be read as an event in this tentative history of portraiture always already on the verge of its own unraveling. As such, this chapter is not about the reversal of biographical erasure in the sense that it would

primarily seek to assess Skloot's attempt at restituting the fragment of a body to the memory of a "personal" whole and its
social world. Instead the perspective is compositional. It even-
tuates in the formulation of a visual assemblage of cells, flesh,
blood, substance, form, and anatomical structure.[9] Here the
object of recognition and restitution is the number of partici-
pants and modes of participation involved in an ongoing process of fragmentation and remembering.

Throughout Skloot's book, the constant back and forth in time
between Henrietta's life, death, and immortality creates points
of passage between human time and cellular time, between the
body object and the body subject, between the history of a body
and the history of a person. The distinction between body as
object and body as subject is not a given. It is the result of framing procedures. Burial and anatomy both frame death, but in
opposite terms: through decomposition, in one case, and dissection, in the other; as a concealed process, in one case, and as a
set of knowable anatomical relations, in the other. The picture
of the body as a set of anatomical relations is the by-product
of a particular framing that is simultaneously spatial (in the
early modern anatomy theater and the modern gross anatomy
lab), instructional (it is integral to the medical school curric-
ulum), technical (relying on dissection procedures, tools, and
skills), and legal (pertaining to the availability of cadavers as
part of forensic investigation or educational purpose). It is
the modern body made legible through the redistribution of
the relationship between the book (the enduring deposit of
recorded, inherited, debated knowledge accumulated over
time), the body (the cadaver decomposing over time), and the
scalpel (or the dissection, understood as a controlled, methodical

decomposition of the body).[10] In *Body of Work* (2008), a memoir
on her experience as a first-year medical student, Christine
Montross reminds her readers that, in the confines of the gross
anatomy lab, the distinction between the body as object and the
body as subject is enacted and reinforced by rituals of anonymity, dress code, a particular light, a particular smell, controlled
access, and constant reference to a dissection manual. The objectification of the cadaver is to be achieved by abstracting the body
from its relational (i.e., emotional, professional, social) network.
A cloth wrapped around the cadaver's head enforces anonymity
and dampens the identification process. But the abstraction is
fragile. Details from the past life of the cadaver as human
subject—chipped lilac nail polish—can trigger identification and
get in the way of the dissection process. Montross recounts how
she experienced this fragility firsthand as a student of anatomy:

> The hands, feet, and head are parts of the body that are instilled
> with character. They can most quickly conjure up an individual
> life. But I cannot take my eyes off the woman's arms. They are
> covered in age spots and thin and long. . . . They are surely
> the arms of an old woman who spent time in the garden or at the
> lake. They are the arms of my grandmother, which I massaged
> for a week before coming to medical school as she lay in bed
> following a stroke.[11]

It does not take much for the anonym body to become relational
again. After all, a biological marker—the age spots—also reads
as biographical marks. And as Christine and her colleagues
slowly and methodically destroy the body on the cold steel table,
it becomes part of her life as a student of anatomy. In *Body of
Work*, dissection and its rituals appear as a disconcerting way to
give respect to the body within the culture of biomedicine.

What a tribute to the biomedical body entails in the context of *The Immortal Life of* Henrietta Lacks becomes legible, I argue, in a scene in which Skloot recounts the reaction of Henrietta's oldest daughter, Deborah, to an image of her deceased mother's chromosomes: "They're beautiful!" she yelled from the porch. "'I never knew they were so pretty!' She walked back inside clutching the picture, her cheeks flushed."[12] The picture is a gift from Christoph Lengauer, a Johns Hopkins University researcher who developed the visualization technique through fluorescence in-situ hybridization (FISH), which made that very image possible. Deborah then brought the image to her younger brother Zakariyya, before they both accepted an invitation to meet with the living HeLa cells in Lengauer's lab and to look at them under a microscope. Deborah's gesture does not open on a scene of recognition, whether it would be the recognition of a biological entity, a body, or a particular form of personhood. Rather, it interrupts the continuum that would have to be hypothesized in an actual scene of recognition between a biological entity, a body, and a form of personhood. The scene exposes the FISH image in its factuality of being an image beyond recognition, although it remains an image of the body; not because the body is what remains a priori, but rather because it is what remains in its capacity as image. What this means is that the molecularization of the human and its body as human body *also* occurs in the history of the relationality between image and body at the point where the history of biology—from anatomy to molecular genomics—bleeds into the history of representation. It bears noticing that with molecular biology we are not dealing solely with concepts of the body (defined by its position in a genealogy, a collectivity, a legal apparatus, a taxonomy—the list goes on) but, contingent on the elements from this expanding list, with the composition and recomposition of what art historian Hans Belting refers to in his study of the emergence of portraiture in early modern Europe as a "portrayable field."[13]

The body portrayed in *The Immortal Life of Henrietta Lacks* is not the body that would be portrayed in a heraldic context: it is not the body positioned by a coat of arms within a genealogical network of deeds and estates as bearer of a social identity. It is not the body portrayed, remembered, and mourned in a family album either. With the medieval escutcheon portrait panels, what is made to endure, Belting explains, is the legitimacy of a line of descent, not the resemblance to an individual or to a body. This type of portraiture exists at the intersection between "the collective body of the family line and the natural body of the living person" (66) as a visual placeholder inserted in the absence of the body of the living person within a network of ceremonial obligations and property rights. The escutcheon stands for a genealogical body and has a body: it can be carried around. While in heraldry the name-bearing body is a medium of genealogical identity, in the humanist portrait the representation of the body is an image of the Self. In the canonical example of Albrecht Dürer's engraving *Erasmus of Rotterdam* (1526) examined by Belting, the printed portrait is a medium of authorial identity (79). The body identified and recognized by heraldry and the humanistic portrait is irrelevant in the context of Homeland Security and its rituals of clearance. Screening has replaced portraiture. Live biometric capture recognizes bodies as moving target, bodies that belong and bodies that do not, bodies that pose a threat and bodies that are threatened. Belting concludes: "There is no concept of the body that is not the product of a specific time and society. An anthropological investigation must take especially careful note of changes in ways of viewing the body and the person, for it is here that perennial questions about what a human being is—in a social, biological, and psychological sense—are posed" (82). Among other untold scenes and scenarios of molecular defacements and survival beyond recognition, the challenge offered by Deborah's gesture holding the image of her mother's chromosomes as the leaving bearer of her name is therefore to develop modes of attending to emergent and resurgent forms of disjuncture between the history of the body and the history of the person.

Paul Rabinow concludes his study of *John Moore v. the Regents of the University of California* with a vision—an animated vision as it is:

> A transformed piece of matter from John Moore now lives forever, reduplicating itself over and over again in jars slowly rotating on racks in a temperature controlled room in Maryland. The cell line is available upon the completion of a form from the requisite institutions and the payment of a nominal fee for handling. These immortalized bits and pieces can then be used to pursue more knowledge, to produce more health, to yield more profit.[14]

Mo cells present a rare genetic mutation whose market value in 1990 was estimated to be over $3 billion. David Golde developed and patented this cell line in 1981 at UCLA Medical Center using bodily matter taken from John Moore while he was seeking treatment for hairy cell leukemia. In 1990, the California Supreme Court ruled against Moore's proprietary rights on his cells and their by-products and against his claim that he had a right to share in the profits they generated.[15] The *tableau vivant* of slowly rotating jars filled with cells is the counterpart of the famous defacement pictured by Michel Foucault at the end of *The Order of Things*:

> Foucault intimated the imminent coming of a new configuration of language about to sweep the figure of Man away like "a face drawn in the sand at the edge of the sea." It now appears that he was wrong: in the ensuing decades language, in its modality as poiesis, has not turned out to be the site of radical formal transformations through which this being, Man, would either disappear . . . or would transmute into a new type of being, as predicted by Gilles Deleuze.[16]

Rabinow's portrait of Moore as a glorious molecular body is suggestive. It identifies biotech, and not the human, as the locus of these transformative forces and the compound of their relation. "One might object, writes Deleuze, that such forces already presuppose man," although Rabinow adds, "in terms of form this is not true. The forces within man presuppose only places, points of industry, a region of the existent."[17] Of course, the human face in the sand at the end of *The Order of Things* is only an image, and the slowly rotating tubes filled with Mo cell lines do not compose a portrait, no more than Langauer's print of HeLa cells do, at least not by the standards of Renaissance portraiture. But, precisely, what kind of images are they? What's left of the relation of identification between body, image, and medium in a context where "there is no conception of the person as a whole underlying these particular technological practices," that is, no "organismic focus" to hold transformative forces and the compound of their relations together?[18]

In his reconstitution of the Moore case's ruling, Rabinow recalls how Stanley Mosk, associate justice of the California Supreme Court at the time, summons in a comment on the case notions of "dignity and sanctity with which we regard the human whole, body as well as mind and soul."[19] Whether or not the notion of dignity is operative in *Moore v. UCLA* without the theologico-political background that medieval historians attached to it and, with regard to the notion of sanctity, whether or not organ transplant technology and the ethico-legal frame that recognizes the status of brain death summons and reorganizes medieval speculations on the relations between the resurrected human whole and its enshrined remaining parts, what matters is how and why their perfunctory summoning obtains in a derived context.[20] Mosk's reference to sanctity and dignity understood as theoretical positions on the relationship between personhood and corporality reintroduces and as such entertains a form of "organismic focus" into a biotechnological field committed to the fragmentation and recomposition of organic matter.[21]

In her study of the Visible Human Project (VHP), Catherine Waldby worked out some of the terms of what would constitute a portrayable field without organismic focus. An initiative of the U.S. National Library of Medicine, the VHP is a digital 3D interactive anatomical atlas completed in 1994 and 1995 for the male and female phases of the project respectively.[22] The VHP performs anatomical legibility through a systematic realignment of volumetric imaging (CT scans and MRI) and cross-sectional photographs (analog and digital) of a prepared cadaver. While the VHP's fifty-nine-year-old female donor remains anonymous, the identity of the male cadaver—Joseph Paul Jernigan, a thirty-eight year old convicted of murder and executed by lethal injection in 1993—was widely publicized. There is in this respect a striking parallel to be observed between Jernigan's virtual life and the double effect of biographical capture and erasure that Foucault describes in "The Lives of Infamous Men" (1977).

This short text was meant to serve as a preface to a project of anthology gathering a selection of internment memos and

letters gleaned in the French penal archives covering the period
from 1660 to 1760. If it had not been for the damning lines
that convicted them for petty crimes or alleged debauchery,
there would not be anything left of these otherwise unremarkable biographical trajectories. There certainly must have been
more to their existence than what the police records show, but
this type of counterfactual is beside the point: the "infamous
men" only have the biography their encounter with the power
structure that terminated their social life gave them. Hence the
logic behind the collection: the texts selected ought to be both
documents and instruments, that is, both the unlikely records
of lives on the verge of anonymity and, at the same time, the
sure signs of their demise:

> What shall be read here is not a collection of portraits: they are
> snares, weapons, cries, gestures, attitudes, ruses, intrigues for
> which the words have been the instruments. Real lives have been
> "played out" in these few sentences; I don't mean by that expres-
> sion that they have been represented there, but that, in fact, their
> liberty, their misfortune, often their death, in any case their des-
> tiny have been, at least partly, therein decided. These discourses
> have really affected lives; these existences have effectively been
> risked and lost in these words.[23]

Commenting on the status of this "verbal existence" on the
verge of fiction,[24] Giorgio Agamben writes: "What suddenly
comes to light is not . . . the subject's face, but rather the dis-
junction between the living being and the speaking being that
marks its empty place. Here life subsists only in the infamy in
which it existed; here a name lives solely in the disgrace that
covered it. And something in this disgrace bears witness to life
beyond all biography."[25] With the VHP, the subject's face and
everything attached to it were deep frozen before being "sliced

into oblivion."[26] Whatever place the body had previously occupied is emptied out as a condition of its being reborn digital, immortal, and fully operable. And yet, as Waldby notices, with the technology that made the VHP possible "the transformation from an every day to an analytic entity has preserved the individuality of each figure in a kind of anatomical portrait, a personal photograph."[27] In her op-ed piece on the sequencing and unauthorized public release of the HeLa's genome, Skloot stresses a comparable point: the identification of Henrietta's family members remains possible.[28] What comes into light with the VHP does not "bear witness to life beyond all biography," as suggested by Agamben, nor does it bear witness to the subject's face beyond portraiture, but rather to a particular mode of biographical capture inflected by a dual logic of erasure and rebirth. Lisa Cartwright explains that

> what is significant is not that Jernigan's identity initially was not known, but that it did become known—moreover, become publicly renowned—at the time that the Visible Man went on-line. It is precisely because of Jernigan's status as less than a private citizen, as a subject stripped of certain rights under the auspices of the state, that he qualified as a universal biomedical subject, and as a public icon of physical health. The universal biomedical subject is thus a subject stripped of his rights to privacy and bodily integrity, even after death.[29]

In that regard, the VHP is more than a digital retooling of anatomy: it is exemplary. It "recapitulates an entire history of anatomy within itself. It is the most recent instantiation of a long biotechnical project, the anatomisation of human cadavers in order to produce the human body as a resource for 'Man.'"[30]

The use of the word *recapitulation,* whose root is the Latin *capitus* (the head), resonates with Skloot's biographical project opposing the retrieval of a metonymic face (Henrietta's)—a face for a life—to a molecular defacement (HeLa)—a fragment of life and life in a fragment. The making of biovalue that produces "the human body as a resource for 'Man'" is headless. It does not need organismic foci or accompanying narratives of personhood and responsibility to generate profits and prospects. The corpse's use-value is indexed in terms of anatomical transparency.[31] But there is also a "recapitulation effect" in the sense that the VHP "produces a 'photorealistic' image, captured in the same spectrum as photography and the human eye, so that the body data appears with the colour values and resolution associated with photographic media."[32] The VHP dataset offers an infinite reservoir of visual possibilities that can be explored through an indefinite number of perspectives and procedures. Anatomical transparency is but a default setting—or heading, hence the recapitulation effect. It imposes limitations on the way the dataset possibilities are used and configured to address the human eye. Unaddressed possibilities were unleashed by Frank Schott and Croix Gagnon to create nonanatomical effects in their photomontage *12:31*—the exact time of Jernigan's execution. The two artists created ectoplasmlike forms by playing the VHP animation "on a portable monitor, which was moved around by an assistant while being photographed in a dark environment. The resulting images are long-exposure 'light paintings' of the entire cadaver."[33] They oppose to the anatomical photorealism the horror of a body with unlimited visual possibilities. *12:31* is nothing but a mishandling of digital information transforming a form of anatomical legibility and transparency coded into the VHP's photorealistic rendition into something like an illegible data cloud. With its fluorescent elongated silhouettes dancing in the dark, the montage thus realizes some of the spectral and choreographic potentialities of the VHP dataset itself.

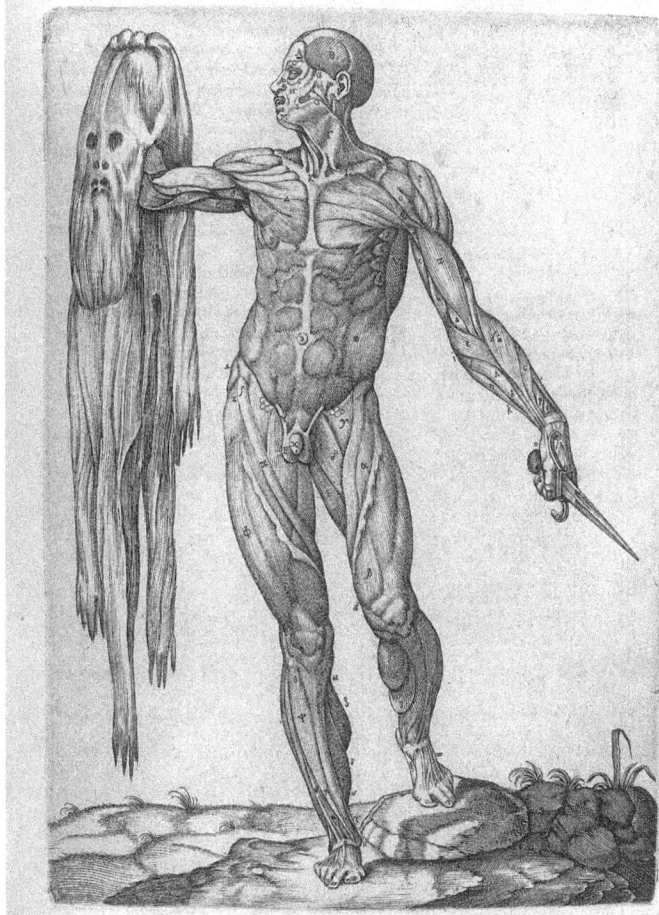

FIGURE 4.1 Juan Valverde de Amusco, *Anatomia del corpo humano* (Rome, 1559), 64. Courtesy of the U.S. National Library of Medicine.

There is a parallel relation between legibility and illegibility to be found in the visual setup of an early modern anatomical figure by Juan Valverde de Amusco. The myology section of his *Anatomia del corpo humano* proposes a variation on the martyrdom of Saint Bartholomew. The skinned figure holds in one hand the instrument of his flaying, and in the other, the peeled-off drape of his skin displayed in a quasi-triumphal gesture reminiscent of Perseus brandishing the severed head of Medusa in Benvenuto Cellini's bronze sculpture on the Piazza della Signoria in Florence or on the antique wall paintings of the Villa San Marco in Stabiae. According to the legend, Perseus triumphed from Medusa's petrifying gaze by tricking the mythological creature into looking at herself in a polished shield. Medusa's images are thus always commemorating both the triumph of representation over the body and the death-bound dimension of images of the body. The two knives and the two cuts—the one that engraves and the one that flays, the one that chisels and the one that slays—are the two sides of the same image-making process. On the one hand, Valverde de Amusco's image of a glorious body made of muscles and legible relations commemorates the triumph of the anatomical gesture over the horror that dissection inspires. On the other hand, the brandished mask of skin commemorates in its own visual terms a body made of illegible relations, a body whose dimensionality escapes the anatomical gaze and the contract on which it depends. As a result, Medusa's ghost in the leftover skin brandished by the flayed figure opens the anatomical image on its visceral undoing in body horror.

If the body is a visual assemblage, if it exists in a visual form and through visual codes, body horror is the corner of visual culture where experiments with these visual properties and codes run wild. Originally, body horror designates a film subgenre that flourished in the seventies and eighties. As Philip Brophy aptly puts it, body horror enacts the body as a visual form with "total disregard for and ignorance of the human body."[34] Body horror occurs at the expense of narration when the visceral takes over narrative continuity, when in front of the splattered screen there is nothing left to be told, or, as in the opening scene of Craig Baker's *Hellraiser* (1983), once we are left to figure out how these bits and pieces of meat scattered or hanging on hooks may have fitted together at some point into a human shape.[35] Back to Agamben's reading of "The Lives of Infamous Men," body horror offers a glimpse at the horror of life beyond biography, of a face beyond its portrait. Or, back to Spillers's American grammar book and her parsing of a syntax of "total objectification," anatomy and body horror are not distinguished in the New World sociopolitical order marked by the "*theft of the body*—a willful and violent . . . severing of the captive body from its motive will, its active desire."[36] Here the discourse on the body and the discourse of torture and enslavement cannot be told apart.

Rather than erasing the past by reforming bodies, biotech interventions create relations of contentious reciprocity between living and nonliving entities, between human and nonhuman entities, and, moreover, between entities living differently in time.[37] These relations of reciprocity in time and across time are instrumental to what biotech is and does if it is not entirely reducible to the description of a process (by which, to take the example of cell nucleus replacement, an oocyte is enucleated and replaced by an adult cell nucleus).[38] It is in this sense that the biotechnological and biomedical present is historical.

In a situation in which FISH ignores the drama of recognition that unfolds through and in front of the visual enhancements it affords, Deborah's gesture confronts the visualizing tool that brings her mother back in bright colors as HeLa. Her embrace both exposes and readjusts relations of correspondence and non-correspondence between parts and wholes, between genetic material and personhood, between her mother's cells and her mother's plight.[39] The scene dramatizes the fact that the technology of tissue culture assigns values to certain bodies and body parts through procedures at odds with the ways family members, loved ones, but also human rights, assign values to lives. The relational space defined by biotechnology pertains to more than bodies and organic matter: it composes, with narratives of destitution, generation, destruction, rebirth, and personhood and its visual expression in the history of portraiture, to form the assemblage that is the immortal life of Henrietta Lacks.

Roland Barthes offers a point of leverage in the lines he devotes to seventeenth-century Dutch guild portraits:

> Entirely identified by their social heredity, these Dutch faces are engaged in none of those visceral adventures which ravage the contenance and expose the body in its momentary destitution [*dénuement*]. What have they to do with the *chronos* of passions? Theirs is the *chronos* of biology; their flesh has no need, in order to exist, to anticipate or to endure events; it is blood which causes it to be and to command recognition; passion would be pointless, it would add nothing to existence.[40]

To be sure, there is an early modern biology of passions historically available to complicate his observations. What I retain from Barthes pertains to the fact that not everything nor everybody will have been alive—and portrayed—in the same terms. The perspective offered by the future perfect is instructive. It highlights the normative structure of representation as a body enters the field of portraiture to play its part in the drama of recognition (whether through straightforward heredity or impossible picaresque plotlines). The distinction between the *chronos* of passion and the *chronos* of biology, between a narrative of transgression and a narrative of transmission, between blood baths and bloodlines, is yet another version of the relation of disjuncture between human time and cellular time.

Skloot opened *The Immortal Life of Henrietta Lacks* with the evocation of the black and white photograph of Henrietta that appears in a corner of the book cover surrounded by the faded image of her immortal and cellular self. In an arresting turn of sentence, the late 1940s photograph of a well-dressed woman is both a portrait of Deborah's mother and an image of her transformation: "Her light brown skin is smooth, her eyes still young and playful, oblivious to the tumor growing inside her—a tumor that would leave her five children motherless and change the future of medicine."[41] Henrietta's photograph is a picture in time and of time in the same way that the VHP remains, despite the properties that a digital interface affords to a mortal body, both a historical artifact and a historiographical machine recapitulating anatomy's violent past. Henrietta's black and white image exists through anticipation and endurance as photograph of Deborah's mother. The FISH rendition of the HeLa chromosomes is an image of time that is not experienced. It is not an "agent of memory."[42] It does not serve as the proof of an existence across the passage of time, only as the site of a disjuncture.

5

ONCOSCRIPTS

When I was first diagnosed with cancer, my initial shock placed me outside the narrative of my own life, watching it as if it were on film. . . . I felt as if someone had got the story wrong: this was not how it was supposed to go. The first reel of the film had finished and the rest of my life, already recorded on the second reel of film that lay next to the projector, ready and waiting to be screened, had been forgotten. The whole episode felt as if it was someone else's script, not mine.

—Jackie Stacey, *Teratologies: A Cultural Study of Cancer*

Not all lives will become biography, autobiography, hagiography, or memoir. What has been described by literary critics as an explosion in the numbers of biographies, autobiographies, and memoirs published each year since the 1980s might suggest that something on the order of a shift has occurred. Conservative commentators describe the phenomenon as an "everything goes" resulting in or from the leveling of genres and values. The line of argument moves in two directions: either it is a change of ethos that translates into an explosion of narratives produced, then sometimes published, sometimes retrieved later on, or, the explosion in the number of narratives produced, published, and retrieved facilitates a change of ethos. Sociologically inflected critique ventures explanations that instead consider the emergence of new existential platforms such as aging, health, and disability in the process of biographical capture.[1] In "The Life of Infamous Men," Foucault already describes a shift: "For a long time, in western society, the life of the everyday could only accede to discourse when traversed and transfigured by the fabulous; it had to be drawn out of itself by heroism, exploit, adventures, providence and grace, possibly, by the heinous crime; it had to be marked with a touch of the impossible."[2] Ordinariness only registered when cautionary or edifying enough. Things changed, Foucault explains, when societies started to "loan words, turns of phrase and constructions, rituals of language to the anonymous mass of people in order that they be able to speak about themselves" (89). When reaching that stage, the problem is no longer to determine which biographical trajectory, fabulous enough or extraordinary enough, deserves to be captured and recounted. To tell the story of postfabulous lives and to picture one's life in a biographical light is less the expression of a privilege or a curse than a duty of expression (*devoir de dire*) that finds a coherence within the history of power as power over the ordinariness of life rather than solely in the form of a repressive state power. According to this model, biographical capture is part of a process that turns the relation "from subject to subject, [relations] between members of the same family, neighborhood relations, relations of commercial interest, of occupation, of rivalry, of hatred and love" into a domain of intervention (85, translation modified).

Arlette Farge, Foucault's former collaborator, observes that "infamy" in "The Life of Infamous Men" is not a code word for homosexuality, as it usually was in the early modern police archives.[3] Infamy designates something like a point of impact. The target was not necessarily the individuals themselves in their exemplarity (as it was the case with extraordinary lives), but the relations that infamous lives entertained with their "passions" and carnal appetites, with the appetites of their neighbors, with excess, with kinship obligations, and with a certain sense of accountability. Infamy pertains to these relations and describes the archival fact that they were deemed subject to ruling and to unruliness. The result is, one the one hand, a figure of destiny that "takes the form of the relation to power, [and] of the struggle along with or against it" (80), and on the other, a corpus of texts that Foucault "resolved . . . to assemble . . . for the sake of the intensity which they appeared . . . to have" (77). The motivation behind the project was not strictly archival. It involved something of the order of affect: "I would find it difficult to say exactly what I felt when I read these fragments and many others which were similar to them. . . . I confess that these '*nouvelles*,' suddenly rising up through two and a half centuries of silence, stirred more fibres in me than what one usually calls literature" (77).[4] In fact, these texts covering the period 1660 to 1760 come from a time in which literary representation was not available in a nonfarcical or noncomical register to the ordinariness of the crimes and deeds they describe. They are not texts transmitted or amenable to transmission either, only the asymptomatic traces of "the working of power on lives and the discourse that is born from it."[5] With *Perishability Fatigue* I want to suggest that the defining traits of destiny are changing as seed banks, genetic engineering, and biomedicine transform the ordinariness of transience into domains of intervention and contestation. The point of impact is our normative relation to loss, storage, and decay but also to the idea of lifespan, understood as a normative formation activated or reactivated by a negative prognosis.

Alison Kafer offers a vivid example of what a shift in the definition of destiny entails from the vantage point of critical disability studies when she writes at the beginning of *Feminist, Queer, Crip*:

> I have never consulted a seer or psychic; I have never asked a fortune-teller for her crystal ball. No one has searched my tea leaves for answers or my stars for omens, and my palms remain unread. But people have been telling my future for years. Of fortune cookies and tarot cards they have no need: my wheelchair, burn scars, and gnarled hands apparently tell them all they need to know. My future is written on my body. . . . What my future did hold, according to my rehabilitation psychologist and my recreation therapist, was long-term psychological therapy. My friends were likely to abandon me, alcoholism and drug addiction loomed on my horizon, and I needed to prepare myself for the futures of pain and isolation brought on by disability.[6]

In this passage, accessibility as political platform in disability right activism takes on a new dimension: what Kafer sees herself denied is not just access to futurity but to a narrative space in which to imagine, plan, and live alternative forms of continuity in time. To live and tell beyond sentencing structures is to resume the story of our verbal existence initiated by Foucault in "The Life of Infamous Men": in the Christian sacrament of penance, the sinner tells everything so that everything in the end can be erased; in the police archives, by contrast, everything that is said will be kept.[7] In the wake of a clinical encounter, prognosis entails a critical relation to a pronouncement and its self-fulfilling properties. A young general internist quoted in Nicholas Christakis foundational study on the subject confides:

> I often had the distinct impression when I talked to patients that I was changing the future. I could never be sure whether presenting what I thought was going to happen would change what would in fact happen. And maybe the way I presented it was important too! So, if I said things one way, one thing would happen, and if I said them another, something else would happen. Because I am convinced that this is the case. The Patient and I were changed by what I said.[8]

A prognosis may very well be "holding together the future and the past" in a statement, but on the patient's side this holding can be corrosive, the opposite of a time of accumulation and flourishing.[9]

Focusing on the cognitive dissonances surrounding cancer culture in the United States, medical anthropologist Sarah Lochlann Jain goes as far as to say that "'living in prognosis' might serve as an alternative to the identity politics that has infused disability studies—and indeed, if pressed, I would argue that all of us in American risk-culture live to some degree in prognosis."[10] Prognosis is a mode of orienting oneself in the meantime. It is a space of representation in which a patient is asked, directly or indirectly, to represent herself in relation to statistical aggregates and position herself in relation to narratives of perishability and survival (beyond five years). It "offers an abstract universal, moving through time at a level of abstraction that its human subjects cannot occupy. . . . Simply a structure of and for our fantasies, the prognosis itself has no time for the human life and death drama" (79). To introduce the notion of oncoscript is to call attention to tools with which cancer culture occupies or reclaims a more or less fragile, more or less normative, more or less accessible form of continuity in time with fantasies, tragedies, cautionary tales or edifying stories, life insurance, life savings, 401(k) plans, clinical trials, fertility treatments, and routine screenings.[11] Oncoscripts belong to this list of tools as a narrative mode of relating to the continuum of possibilities that cancer diagnosis and prognosis disrupt, but within which cancer also exists simultaneously as a disease, a statistical event, a death sentence, a turning or tipping point in someone's life, a source of pharmaceutical profits, a leading cause of bankruptcy, and a biographical capture or entrapment.

FIRST SCRIPT

The point of departure is not the self, as it is the case in Rita Charon's vision of narrative medicine, where "telling and listening to stories seem as organically necessary as are the respiration of oxygen and the circulation of blood to establish and maintain a self by metabolizing into it that which is nonself and then contributing products of the self back into that alien domain, thereby making it home."[12] At the origin of an onco-script there is the effect of disjuncture that detection and screening practices open between the biomedical body and a lived experience of illness.

In *The Normal and the Pathological*, Georges Canguilhem imagines the following forensic paradox based on the theories of French surgeon René Leriche (1879–1955):

> If an autopsy of medical-legal intent were to reveal a cancer of the kidney unknown to its late owner, one should conclude in favor of a disease, although there would be no one to whom to

attribute it—neither to the cadaver which is no longer competent, nor retroactively to the formerly live man who had no idea of it, having had his life come to an end before the cancer's stage of development at which, in all clinical probability, pain would have finally announced the illness.[13]

Here we do not have a cancer narrative per se but an oncoscript. The notion of illness narrative would have been foreign to Leriche's hypothetical case. There is nothing in his clinical practice to recognize that "the patient is somehow both the object of medical work and an object of speech (spoken to and about)."[14] By the second half of the twentieth century, the notion of illness narrative had gained considerable traction, at least in the U.S.[15] The scholarly response to the editorial proliferation of illness narratives and to their occasional dismissal as "victim art" shows that illness narratives are not just an anecdotal archive but a medium for the politicization of patienthood with its own trajectory within the history of public health and medicine. In her extensive study of published illness narratives, Lisa Diedrich identified a shift, for instance, from patient activism, calling for the transformation of society and medical institutions, to a neoliberal focus on personal responsibility and self-transformation.[16] With Leriche's forensic paradox, there are no narrative provisions that would allow the patient to undertake a series of corporeal and spiritual transformations. The layering of tissues does not add up to form something like a biographical entity. Cancer was there before the narrative that strives to situate it as an event on a biographical timeline. A biopsy can give an estimate of when it all started, but this beginning is not articulated as such, for cancer inception is not discursive but cellular. The description of this biochemical process is but a model and an approximation.

Canguilhem adopts a different outlook on Leriche's paradox. He reframes the deductive contours of the hypothetical case by foregrounding the temporal and experiential realities that inform histopathology's insights. "If, today, Canguilhem argues, the physician's knowledge of disease can anticipate the sick's man's experience of it, it is because at one time this experience gave rise to, summoned up, that knowledge."[17] In other words, Leriche's unattributable cancer is but an isolated episode in a narrative arc. Catherine Belling alludes to the same logics when she notes that, "the discovery of cancer does not emerge from a narrative vacuum. In order to be diagnosed, it must already be present both in the patient's body and in the cultural imaginative landscape of patient and clinician (and, hence, society)."[18] It does not mean, as she explains later in the same article, that cancer diagnosis has not been envisioned in postclinical terms as a "rule-based process identified not by its emergence in any particular organ or at any site in a body's geography, but by its dispersal, throughout all living cells, as one generative possibility among many." Belling pursues: "We would need to invent new patient narratives, fundamentally postmodern ones, to tell about such cancer" (254). As the biography of a cell line derived from a cervical cancer that not only outlived but also exceeded in volume the body in which the initial tissues were harvested, Skloot's project in *The Immortal Life of Henrietta Lacks* already belongs to this next generation of cancer narratives.

SECOND SCRIPT

For American author and activist Barbara Ehrenreich, cancer is simultaneously a histological reality legible in the pathology slides from her breast biopsy and a grammatical event in the form of a diagnostic sentence pronounced by the surgeon: "Unfortunately, there is cancer." She recounts in her influential essay "Welcome to Cancerland":

> It takes me all the rest of that drug-addled day to decide that the most heinous thing about that sentence is not the presence of cancer but the absence of me—for I, Barbara do not enter into it even as a location, a geographical reference point. Where I once was—not a commanding presence perhaps but nonetheless a standard assemblage of flesh and words and gesture—"there is cancer." I have been replaced by it, is the surgeon's implication. This is what I am now, medically speaking.[19]

From now on, in the aftermath of the biopsy results, she still is, but in a timeline indexed on the diagnostic sentence "there is cancer." In Ehrenreich's parsing of the new biographical configuration she finds herself in, it is as if the present of cancer was incompatible—noncontinuous—with the past where "I once was." Whatever "assemblage of flesh and words and gesture" persists or perishes in the wake of who "I" once was, her oncoscript exists as a reminder that a diagnostic statement is not the truth that will save the subject.[20]

THIRD SCRIPT

Kathleen Woodward isolates the following script in the 1990s medical drama *Chicago Hope* (episode 21, season 3):

> A middle-aged woman—she is a wife and mother of two children—insists to a young male surgeon that she wants a double mastectomy. He is not only reluctant to do the operation, he is

horrified, because she does not in fact have breast cancer. But as she explains, she has an eighty-six percent chance of getting breast cancer. . . . In the end [the doctor] is persuaded by her unwavering determination and the gravity of her statistical prognosis to perform the operation. . . . What to the surgeon at first seems an insane course of action is revealed in the course of the narrative as preeminently rational in an unequivocally calculating sense.[21]

In this episode two things became indistinguishable: while renouncing one's breasts is presented as the logical, even responsible, thing to do, adopting a rational behavior amounts to renouncing part of oneself.

Precancer, however, is not felt. It is diagnosed. A patient does not feel precancerous. She has to wait for the results of a biopsy or the results of a screening for mutations in the BRCA gene that put women at a higher probability of developing breast or ovarian cancer.[22] As a practice, preventive mastectomy turns the disjuncture between the biomedical body and the lived experience of illness into a scripted site of intervention and self-reflection. As a scenario, preventive mastectomy is both informed by an ascetic rationalization of the ontological uncertainty regarding the fact of having no more than a shot at the biological history of life and scripted as a form of resistance to the disembodying experience of invasive cancer.

For historian Barbara Duden, cancer is disembodying, but on different grounds. In her 1997 address to the German Cancer Society (Deutsche Krebsgesellschaft), she is adamant: tumors have existed in the past, and oncological categories are already recognizable in the Hippocratic corpus (400 BC), but the association between cancer, prevention, and risk is "foreign to the past."[23] Body history, understood as the history of how individuals sensed themselves, is, for Duden, the script of her resistance to "disembodying rituals" that target through probabilistic calculation, "even in these curves and recesses of the body, that were traditionally understood as places of a more intense life-giving aliveness, and as places of desire, pleasure, and longing."[24] In her own words, body history is a form of *ascesis*, that is, a disciplined reflection and intervention rescinding the surgical and statistical rationalization of viability in contemporary oncological practices.

Reflecting on her own clinical encounter with a cancer diagnosis and the statistical existential template of cancer survivorship, Lochlann Jain orients her rationalization of viability toward the task of learning how to live in prognosis. "Living in prognosis," she explains, "severs the idea of a timeline and the usual ways we orient ourselves in time: age, generation, and stage in the assumed lifespan."[25] To live in prognosis is to let go of scripted biographical trajectories. It is to find oneself "outside the narrative of [one's] own life" (Stacey in the epigraph), having to improvise nonlinear prospects of continuity in time, without relying on principles of accumulation over time. In other words: "If you are going to die at forty, shouldn't you be able to get the senior discount at the movies when you're thirty-five?"[26] With or without a discount, the oncological significance of Agnès Varda's film *Cléo de 5 à 7* (France, 1962) speaks to this task.

FOURTH SCRIPT

June 21, 1961, from 5 to 7 P.M., a little less than two continuous hours in the life of Cléo are divided into thirteen chapters and stretched between two prognostic statements. A tarot card reader delivers the first prognosis, not as a deadline but in the more spiritual terms of a "profound transformation of [her] whole being." The second and final prognosis takes the form of a radiation therapy referral delivered nonchalantly by her doctor. In both instances, cancer is not named. Cléo has already left when the tarot card reader declares: "The cards say death, and I saw cancer. She is doomed." Meanwhile, Cléo enters prognostic time. With or without a clear diagnosis, she is persuaded that there is something wrong with her even though none of the many mirrors punctuating her journey through the streets of Paris can give a form to her sense of distress. She cannot be pretty and sick at the same time. Catching one of her reflections in a mirror, Cléo muses in a voice-over: "This doll's face always the same. . . . I cannot even read my own fear in it. I always thought everyone looked at me, and I only look at myself." Here cancer is not a lesion or even a pathological event. It belongs to a cinematic trajectory: it is a variable in Cléo's radical transformation from being a woman looked at to being "an active social participant, rupturing the oppressive unity of identity and vision and appropriating the gaze for herself in a new appreciation of others in the world around her."[27]

Cléo is by no means a medical film. Cancer is not, in Varda's film, a biological reality. It is a word Cléo hears and does not get to hear; it is something she fears at first and then, by the end of the film, does not seem to fear anymore. Cancer is not, for her, an oncological reality, only a feeling. Cléo's disease exists solely at the level of perception. It is a nauseous episode or an overwhelming sense of fatigue and anxiety. It is part of a sensory vocabulary that pertains to the positioning of her body on the screen ("everyone looked at me, and I only look at myself"). When Dorothée, Cléo's friend, inquires, "Where is your disease?" She answers, "My belly. I prefer to have it there than somewhere else. . . . At least you can't see it," meaning it is yet to be a public platform of mobilization. As such, Cléo does not inhabit a landscape of risk.[28] She is not presented as a member of an ever-evolving statistical population in which a certain percentage of women will be diagnosed with cancer at some point during their lifetimes. In the film, cancer is not caused or linked to environmental factors. It is not inherited and part of an intergenerational plotline (as in a movie like *Five*).[29] Cléo's fear is not legible against a background of cancer prevalence and of survival rates. Conversely, her oncoscape is not an abstract space of statistical or surgical possibilities. It is urban, cinematic, and sensorial. It is shot through and through by the pulse of "ordinary affects," humming radiophonic voices, familiar and unfamiliar faces, eavesdropped conversations, and inquisitive gazes.[30] It is coterminous with the motion of lights through paulownia foliage on the Place d'Italie. It encompasses Cléo's continuum of accelerations and decelerations through Paris by car, taxi, tramway, or on foot.

FIFTH SCRIPT

In his study of Antonello da Messina's *Saint Sebastian* (circa 1450), art historian Daniel Arasse distinguishes between two bodily contours: the body of the saint in painting (*le corps en peinture*), sitting within the perspectival grid and recognizable in its anatomical features, and the body of painting (*le corps de peinture*), a devotional assemblage of gestures and pigments.[31] The distinction is theoretical insofar as the coincidence between the two contours can be more or less visible and appears more or less seamless. Precisely, Arasse locates a point of disjuncture between Sebastian's two bodies at the level of the saint's navel.

The anatomical detail is not where it should be. It is slightly off-center both with respect to an idealized body image and with respect to the perspectival grid in which the painted body is assembled.

In the hagiographic tradition, Sebastian's relics and the depictions of his martyrdom offer a protection against the assault of the black plague represented by the arrows that pierce his flesh but fail to kill him. They are also known to be an efficient cure against concupiscence. Antonello makes room for Sebastian's body in a way that features his unsullied skin. The few arrows that reached him went through the flesh without causing much damage. Moreover, this *Sebastian* puts his spectators in the delicate position of being either the archers taking aim at the saint's flesh or being aroused by it. But both scenarios, whether they consider the body as a target or in its tempting nudity, are stained by the displaced navel of the saint. The displaced anatomical detail is now but a stain (*macula*) on the body in painting and a pictorial aberration. It is the symptom of a disturbance in the relation to an image of the body targeted, desired, adored, or addressed.

We are as far from oncology and cancer culture as might be. Antonello's painting is not about cancer; the diagnosed *macula* has indeed no clinical bearing. There is nothing oncological in this oncoscript. The stain is not a lump, a mass, a tumor, that is, *onkos* in Greek. It is not the growth of a site of inflammation or one that would show up on an MRI or leave a chemical trace in a blood test. It points at the hardening of the body into legible and illegible signs. Likewise, when, in Duden's work on early modern gynecology and obstetrics, eighteenth-century women from eastern Germany complained about inner ebbing, flowing, curdling, clotting, disorderly fluxes, or spoke about the imbalance of their humors, for them and their doctor, there is nothing to see: "most of these clots are nothing which can be fathomed by magnetic resonance, ultrasound echo, or X ray."[32] Rather, they pertain to "a way of being, feeling and sitting within oneself that is oriented not primarily by visual reference but by touch, taste, the sense of space, the feel for atmosphere."[33] Their lumps are not histological or physiological entities, which is not the same thing as saying that they did not exist in the particular clinical encounter that shaped them.

SIXTH SCRIPT

With the installations and sculptural projects of Leonor Caraballo and Abou Farman, we are immersed in contemporary cancer culture, understood as "a space that provides possibilities for Emergence, even in, and perhaps most especially in, a context that otherwise would be one entirely organized around cancer as Emergency."[34] Started in 2010, *Object Breast Cancer* is a series of three-dimensional renditions, or "extractions," of breast tumors based on Caraballo's own MRI scans. Cast in bronze, the tumor becomes an object to be experienced in visual terms. Smaller casts of the tumors can be worn like pendants or amulets, sitting on the surface of the body rather than lodged in its insides. With them, breast cancer becomes public in a color other than pink. Caraballo and Farman sculptures make tumors visible in two ways: to the patient herself, and to others, as potentially confrontational objects and disconcerting conversational pieces, but also as medical specimens offering diagnostic and treatment vantage points on tumors, or as offerings.

There is a votive dimension to *Object Breast Cancer.* The organic forms that "extractions" generate would not be out of place if found next to anatomical ex-votos: arms, ears, lungs, legs, and other ailing organs fashioned in clay or in wax, described by art historian and specialist of visual culture Georges Didi-Huberman:

> Someone who suffers in the right side of their chest will dedicate an ex-voto representing only the right side of a bust. And if they suffer to the very depths of their lungs or their guts, they will not hesitate to sculpt organic forms of them, half-observed and half-imagined. The anatomical ex-voto thus presents itself as a fragment, cut out in accordance with the fault-lines of the symptom itself. Its actual size often takes on the meaning of a survival or of an imitation of the protocols of the imprint, the suffering organ being, if possible, directly moulded in order to be devoted with greater precision and auratic intensity.[35]

The parallel between the relics engineered by Caraballo and Farman and the ex-voto tradition runs even deeper. It foregrounds the conditions in which Caraballo's tumor gained access to representation as an image of vulnerability rather than as an oncological reality. Indeed, "before representing anyone, the ex-voto represents someone's symptom and prayer: what the giver models in wax is primarily *the site of [her] suffering* and *where [she] wants to be transformed*, soothed, healed, converted."[36] It is this votive dimension of diagnostic technologies depositing an image of the body that *Object Breast Cancer* also captures. It is this dimensionality of magnetic resonance imaging that Donna Haraway captions in Lynn Randolph's oil painting *Immeasurable Results* (1994): "the moment of reading and scanning, of being read and being scanned, is the moment of vulnerability through which new articulations are made."[37] The resulting image is one of these emerging articulations—and so is life in prognosis. There is clearly something wrong with the transcranial scan that appears in *Immeasurable Results*: "A fantasy mermaid with an open fish mouth: a parallel floating penis and testes of the same piscine shape as the doll's: a pocketwatch without clock hands, armed instead with crab claws . . . a red demon hammering at the skull."[38] These dancing growths and creeping lesions are not legible in the color-coded calibration of tissular masses. They are, to return to Didi-Huberman's description, the "relief of a psychically processed organic ordeal."[39]

Unlike ex-votos, *Object Breast Cancer*'s extractions are not deposited in a sanctuary. The extraction process—the generation of MRI scans, the digital rendition of the tumor, the bronze casts, and the exhibit setup—magnifies the cancer diagnostic and projects it in space: in a space that might very well be filled with known or unknown or unacknowledged carcinogenic substances. In other words, Caraballo and Farman's *Extractions* are counterextractions, extracting cancer from its societal and political outside, from the nonplace of contradictions in which "cancer definitions clog."[40] It is an extraction from the strategies of cultural, cosmetic, and corporatist containment described by Lochlann Jain.

Another way to understand the geometry of Caraballo and Farman's intervention is to return to Foucault's description of the changing relations between diagnosis and the lived experience of illness in *The Birth of the Clinic*. As an injunction, that is, as a call for objection rather than acquiescence to existing models—and thus in line with Ehrenreich's own objections to a certain form of cancer culture ("Let me die of anything but suffocation by the pink sticky sentiment embodied in that teddy bear"), *Object Breast Cancer* reenvisions the clinical "superposition of the 'body' of the disease and the body of the sick man."[41] This superposition, writes Foucault, "is self-evident only for us, or rather, we are only just beginning to detach ourselves from it." Reconfigurations of this relation happened before, and are likely to happen again. Clinical medicine—in which "the human body defines . . . the space of origin and of distribution of disease"[42]—is the third configuration identified by Foucault. In the first one, the body of the disease corresponds to a surface of classification—the table of symptoms rearticulating dispersed symptoms into a coherent condition with a name and a treatment. The second is sociopolitical and defined by state institutions monitoring epidemics. Here the body of the disease finds its coherence in the language of statistics. This series of shifts obviously does not mean that people did not get sick, recover, or die. People, and not tables nor statistical deviances, experienced illness before clinical medicine, but what they experienced could not register as such (in the same way that certain forms of literary representation were not available to the tragedy of ordinary life in the *Ancien Régime*). Conversely, in the clinical model, and as we have already seen with Ehrenreich's account, a diagnostic sentence ("There is cancer") can prevent a grammatical possibility ("I am") from finding a point of insertion in "a standard assemblage of flesh and words and gesture."[43] The notion of oncoscript signals potential moments of detachment from the self-evidence of the superposition of the body of the disease and the diseased body. It points to the reassemblage of new forms of evidence, but does not identify, nor does it seek to identify, something of the order of a shift in the grounding of medical knowledge. More modestly, this succession of six short scenarios gestures toward the meantime in which moments of detachment are susceptible to be endured in emerging biographical terms.

6

DISPATCH FROM THE PALLIATIVE PRESENT

In the next place, I attentively examined what I was and as I observed that I could suppose that I had no body, and that there was no world nor any place in which I might be; but that I could not therefore suppose that I was not; and that, on the contrary, from the very circumstance that I thought to doubt of the truth of other things, it most clearly and certainly followed that I was; while, on the other hand, if I had only ceased to think, although all the other objects which I had ever imagined had been in reality existent, I would have had no reason to believe that I existed; I thence concluded that I was a substance whose whole essence or nature consists only in thinking, and which, that it may exist, has need of no place, nor is dependent on any material thing; so that "I," that is to say, the mind by which I am what I am, is wholly distinct from the body, and is even more easily known than the latter, and is such, that although the latter were not, it would still continue to be all that it is.

—René Descartes, *Discourse on the Method* (1637)

In the last chapter, the meantime of perishability corresponds to what I propose to call the palliative present.
French linguist Emile Benveniste explains that, "no matter
what the type of language, there is everywhere to be observed a
certain linguistic organization of the notion of time. It matters
little whether this notion is marked in the inflection of the verb
or by words of other classes. . . . In one way or another, a language always makes a distinction of 'tenses.'"[1] How does this
translate in the clinical context of palliative care? Based on her
fieldwork conducted in the early 1980s, Roma Chatterji notes,
for instance, that in the *verpleeghuis* (Dutch nursing home)
"time [is] oriented to the present, since the future could be perceived only as a kind of loss."[2] There is a "grammar" of tenses in
the *verpleeghuis* that corresponds to an observed configuration
of continuity in time. The distinction between past, present,
and future is not primarily marked by the inflection of verbs or
by other classes of words, but in the institutional commitment
to certain practices of caregiving and the emergence of a "new
self-consciousness about the dying process."[3]

By contrast, for experimental filmmaker and cultural critic
Eric Cazdyn, the palliative is less a countdown of the time that
is left after the failure of curative medicine than a radical way to
"reopen the future as something more than a mere linear extension of the present"—as something else than a "forever sick" (5).[4]
The time of the palliative is to be understood in relation to a
biomedical present that turns terminal illnesses into chronic
conditions so that "the patient is now afforded a meantime that
functions like a hole in time" (4). Cazdyn gives a distinctly
political dimension to this idea by opposing terminality to what
he describes in *The Already Dead* as the new chronic: a present
that has given up on the ideas of revolution and cure altogether.
In this configuration of tenses, the perception of loss changes as
a result: "death is the pure form of radical change, and once our

deaths are taken away from us in the name of the chronic then so is our capacity to imagine other radical possibilities, such as cure and revolution" (7). However, this statement is complicated by the fact that the author found himself having to live in this present as the beneficiary of a pharmaceutical product (a tyrosine-kinase inhibitor) that transformed a terminal diagnosis (chronic myelogenous leukemia) into a manageable condition. By anchoring the history of the biomedical present in the retelling of his cancer survivorship, Cazdyn gives a biographical coherence to the duration of a chronic present embedded in a technological state of affairs.

As for what comes after the chronic present, and from the perspective of tense differentiation, palliative time opens on a conception of the biomedical future that differs from the Cartesian prospect that, one day, "we might be freed from innumerable diseases, both of the body and of the mind, and perhaps even from the infirmity of age."[5] Eric Krakauer, physician at the Massachusetts General Hospital, defines palliative care as an ethical response to that version of the future, or, in more programmatic terms: "Palliative medicine was provoked or called into being by the suffering that the Cartesian medical project ignored, produced, and reproduced" (390). But Krakauer is also careful to divest his views from technophobic stances: "Because it is a response to suffering, palliative medicine endeavors to use technology responsibly. . . . [It] decides on use of technologies by hearkening to the voice of suffering and by responding to its call" (391). Here palliative care exists as a domain of counteroperation that puts at stake what it means to envision and cultivate possibilities—including communicational possibilities—in the midst of the terminal.

This takes us to the following projection regarding the palliative future based on an article published in the *New England Journal of Medicine* studying the "willful modulation of brain activity in disorders of consciousness." The authors suggest that "functional magnetic resonance imaging, or fMRI, might someday be used as a communication tool for patients . . . in the vegetative and minimally conscious states."[6] In this scenario an advanced technology produces legibility where there was none; it embeds a palliative future in biomedical promises. The object of focus is brain activity, but the possibilities being envisioned are dialogical. The research effort is geared toward prolonging communication and monitoring of pain.[7] If we were to understand the scope of palliative care in such terms, its domain would be less defined by the form of a pathology, whether chronic, degenerative, or terminal, than by a demand or a complaint waiting to be heard and answered by caregivers. In this model, communication, even in the absence of a voice and in the absence of a call, is both the horizon and the problem of palliative care, not death. For death no longer has a meaning in the biomedical present—"It has none," writes Max Weber in 1917, "because the individual life of civilized man (*Kulturmensch*), placed into an infinite 'progress,' according to its own imminent meaning should never come to an end; for there is always a further step ahead of one who stands in the march of progress." In this configuration of tenses, modern patients—unlike "Abraham, or some peasant of the past," who could die "satiated with life" (*Lebensgesättigt*)—die "tired of life" (*Lebensmüde*), or sedated, having lost consciousness and the ability to say that they have had enough.[8]

The death of *Kulturmensch* is the result of a physiological process of exhaustion, and physiology becomes, in turn, a form of eloquence captioning the progressive loss of vital functions. It is the eloquence deployed by Xavier Bichat, who writes in his *Recherches physiologiques sur la vie et la mort* (1800): "Consider man, who fades away at the end of a long period of old age. He dies in details: one after another, his external functions come to an end; all his senses successively shut down; the ordinary causes of sensations no longer leave any impression on them."[9] This exhausted body dying in details is at the center of Allan King's actuality drama *Dying at Grace* (2003, Canada). In the DVD audio commentaries, King insists that the documentary is not about palliative care, but rather the experience of dying shared by five terminally ill cancer patients, "in the hope that it would be useful to the living." These are, quoting from the opening credits, the terms of a contract in which Carmela, Joyce, Rick, Lloyd, and Eda agreed to be accompanied in their last moments at the Toronto Grace Health Centre. Even if the documentary does not pledge to alleviate the suffering it records, there is an implicit palliative contract that subtends *Dying at Grace*. I use the word *contract* by reference to the encounter between eighteenth-century political theory and the organization of teaching hospitals in postrevolutionary France. Foucault reconstitutes the terms of this contract in *The Birth of the Clinic*:

> Can pain be a spectacle? Not only can it be, but it must be, by virtue of a subtle right that resides in the fact that no one is alone, the poor man less so than others, since he can obtain assistance only through the mediation of the rich. Since disease can be cured only if others intervene with their knowledge, their resources, their pity, since a patient can be cured only in society, it is just that the illnesses of some should be transformed into the experience of others; and that pain should be enabled to manifest itself.[10]

In *Dying at Grace* the spectacle of pain mobilizes palliative medicine. Pain is the object rather than a pathway.

The clinical captioning provided by nurses in their audiotape records is only one verbal thread in a soundscape of respiratory machines and labored breathing punctuated by whispered prayers, words of comfort, and spoken fragments of a patient's biography: Carmela is a first-generation Italian immigrant with advanced ovarian cancer. Though she remains almost completely silent in front of the camera, her relatives and family friends share their memories of a dedicated woman. Joyce is Carmela's roommate. She has lost most of her family, including her children. Rick is an ex-biker with a history of substance abuse. Lloyd is a pastor and the youngest of the patients. Eda's state is improving to the point that she decides to look for an apartment in the city and regain her independence. The documentary, however, ends with the uncaptioned recording of her last breath. In terms of cinematic time, the terminal is more than a prognostic category that identifies a physiological spectacle. It is a mode of editing the physiological exhaustion of bodies captured over the course of several months in a 148-minute feature film. From the perspective of tense differentiation, the terminal is a state of sentience with its own repertoire of repeated gestures and tropes, with an eloquence of its own: checking on a patient, taking a turn for the worse, getting on with one's life, going out for dinner, and letting go.

The focus on communication in the domain of palliative care corresponds to a shift in the fantasy of the good death, from "a sudden and unconscious death (typified by dying in one's sleep) to an aware death that is individual-specific, that is subject to individual control, and that allows the patient to finish business (dying 'my way')."[11] This recentering is legible in Jean-Christophe Mino and Emmanuel Fournier's title *Les Mots des derniers soins* (The words of terminal care), asking: "What do we say to the patient at this ultimate stage, where the disease resists and ends up "slipping through" [*échapper*], or when curative care has to become palliative care?"[12] In the last chapter of a book mostly known in the U.S. for its contribution to the then nascent

discipline of cultural studies rather than for its views on palliative care, Michel de Certeau shares similar concerns: "Around
the dying man, the staff of a hospital withdraws. . . . This dis-
tancing is accompanied by orders in a vocabulary that treats the
patient as though he were already dead. . . . The dying man
ought to remain calm and to rest. Beyond the care and the seda-
tives needed by the patient, this order appeals to the people in
its surrounding's inability to *bear the uttering* of anguish, despair,
or pain."[13] The focus on communication is legible too in this
passage from *The Practice of Everyday Life*, but in reverse. Here
communication is an issue, not a challenge. Because in this
configuration of the terminal the patient is already dead,
communication—verbal or reduced to the expression of
anguish, despair, or pain—is unbearable but not necessarily
impossible; hence the preemptive movement of withdrawal.
Should the already dead speak and be heard, it would "dis-
turb . . . ward routine and turn . . . attention to the moral
dimensions of illness, suffering, and the fear of death."[14] It
would disrupt a therapeutic "'scenario' constructed by a vision
of the future that seeks to make the body what a society can
write." In this scenario, de Certeau pursues, "the relationship of
life to what is written has gradually taken on, from demography
to biology, a scientific form whose postulate is in every case the
struggle against aging considered sometimes as an inevitable
fate, sometimes as a set of manipulable factors."[15] The emergence of palliative care as a clinical matter of concern, and later
as an area of specialization focusing on the alleviation of pain
and quality of life in terminal patients, registers a statistical
trend: death tends not to happen at home anymore. The term
palliative, first introduced in English in the early 1970s by
Canadian urologist Balfour Mount, following the work of Cic-
ely Saunders in England, does not appear in *The Practice of*

Everyday Life, which is the result of research conducted between 1974 and 1978. To name the palliative in "The Unnamable" is to think along a different timeline that has perhaps more to do with tense differentiation than with historical shifts. It is also to recognize that something else than communication can occur in language.

Archaeologists will remind us that the already dead do speak. Ancient Greek funeral objects, for instance, bear inscriptions such as "Eumares built me as a monument."[16] For philologist Jesper Svenbro, "These inscriptions are not transcriptions of something that could have been said in an oral situation and subsequently transcribed upon the object. . . . Quite to the contrary: these statements are in some sense characteristic of writing, which allows written objects to designate themselves by the first person pronoun, despite the fact that they are objects and not living, thinking beings gifted with speech."[17] "I write" is the surrogate for the impossible sentence "I am dead." The archaeological detour brings to the fore, not necessarily current practices in palliative care, but the reconfiguring agency of the "I write" in a text marked both by a form of physiological exhaustion and the search for a certain eloquence.

In Jean-Dominique Bauby's memoir *The Diving Bell and the Butterfly*, published only days before his death in March 1997, the former editor of a well-known French lifestyle magazine silently spells out the terms of his metamorphosis:

> My life toppled over Friday, the eighth of December, last year.
> Up until then, I had never even heard about the brain stem. I have since learned that it is an essential component of our internal computer, the inseparable link between the brain and the

spinal cord. I was brutally introduced to this vital piece of anatomy when a cerebrovascular accident took my brain stem out of
action. In the past, it was known as a "delirium" [*transport au
cerveau*] and you simply died. But improved resuscitation techniques have now prolonged and refined the agony. You survive, but
you survive with what is so aptly known as "locked in syndrome."
Paralyzed from head to toe, the patient, his mind intact, is impris-
oned inside his own body, unable to speak or move. In my case,
blinking my left eyelid is my only means of communication.[18]

Being conscious after the cerebrovascular accident means an
awakening of the subject to an estranged body, even though it is
a body whose physiological functions appear in a clearer light to
the patient. Catching a glimpse at something unrecognizable
reflected in a glass case, Bauby relates: "I saw the head of a man
who seemed to have emerged from a vat of formaldehyde. His
mouth was twisted, his nose damaged, his hair tousled, his gaze
full of fear. One eye was sewn shut, the other goggled like the
doomed eye of Cain. For a moment I stare at that dilated pupil,
before I realized it was only mine" (25). It is certainly symp-
tomatic in this scene, at least in the English translation, to see
Bauby's image appearing recognizable to himself as if it had
been birthed by an embalming fluid (the formaldehyde). In
French the vat of formaldehyde is a barrel of dioxin (*un tonneau
de dioxine*). The passive construction of the original sentence
("Dans un reflet de la vitrine est apparu le visage d'un homme")[19]
does without the "I," leaving only a reflected image in the glass
case. *Vitrine* hints at the world of preserved medical specimens.
If this unrecognizable body is inhabited, it is on a confining
mode (like a diver worn by his diving bell). Once it has been
recognized, it is as a "flabby disarticulated body, which only
belonged to [him] anymore to cause [him] pain" ("corps flasque

et désarticulé qui ne [lui]appartenait plus que pour [le] faire souffrir"). If something in *The Diving Bell and the Butterfly* sounds like a Cartesian thought experiment gone wrong, it is only by overlooking the fact that the thinking subject is a patient surrounded by a team of nurses, speech therapists, physicians, and visitors. To the immediacy of the Cartesian "I think" corresponds the surrogacy of the "I write" in a relation of caregiving. As such, the impetus behind Bauby's text is not epitaphic. It is both life-affirming and epistolary. The chapter titled "The Vegetable" ("*Le légume*") reveals that Bauby conceived his text as part of a long letter sent to friends and loved ones to repair the damage caused by the rumor of his being in a vegetative state: "Thus was born a collective correspondence that keeps me in touch with those I love" (82). Echoing the "not dead yet" of disability activism, *The Diving Bell and the Butterfly* refuses the erasure of social death.

To "write," Bauby used a rearranged alphabet (E S A R I N T . . . etc.) and one of his eyelids as a blinking cursor: "It is a simple enough system. Someone reads off the alphabet in the ESA version until, with a blink of my eye, I stop my interlocutor at the letter to be noted" (20, translation modified). Selected letters come together to form words and sentences. The text is rehearsed as an inner stream of consciousness before being transcribed by Bauby's assistant Claude Mendibil. The collaborative writing sessions generate text, but they are equally time spent with someone rather than alone.[20] The text generated mirrors the gestures of his physical therapist, creating points of contact between massaged and messaging surfaces: "With her warm fingers, Brigitte travels all over my face, including the barren zone which to me seems to have the texture of a parchment, and the area that still has feeling, where I can still frown an eyebrow" (16, translation modified). The text thus exists as a form of surrogate continuity between seemingly incompatible bodily states—the body before and after the stroke, between what is reparable and beyond repair, consciousness between confidence and despair. The compositional effort that anchors Bauby's rediscovery of sensory landscapes and forms (the pink hue of the bricks after the rain, voices heard over the phone, assaulting noise of the hospital, the smell of fries . . .) is on par with the rehabilitation process itself. These memories compose the sensorial vocabulary of a present of reduced mobility. Conversely, the process of physical rehabilitation is a form of remembering that creates new articulations between past and present, but also between locked-in syndrome infancy and convalescence ("I can find it amusing in my forty-fifth year, to be cleaned up and turned over, to have my bottom wiped and swaddled like a newborn's," 16), between father and son—that is, between the reduced confinement of Bauby's aging father in his apartment and his own immobilization—or between his forced immobility and the acrobatic agility of his daughter.

In the 2007 filmic adaptation of Bauby's text by Julian Schnabel, the situation of surrogacy between the object (the written text) and the first person hinges less on the impossibility of saying "I am dead" than on the utterance of a death wish fulfilling the possibility of speaking and dying at once.[21] Departing from Bauby's memoir, Schnabel imagines a scene in which Bauby's reentry into language starts with the painstakingly composed sentence *Je veux mourir* ("I want to die"), or *Je veux mour*, to be exact. His speech therapist completes the two missing letters, lending her voice to the assemblage of syllables. Literally bearing the utterance of his anguish, the speech therapist finds herself in the position of the inscribed archaeological object in Svenbro's model, albeit a sentient one wishing herself out of an exchange beyond communication. She storms out of Bauby's room, leaving behind an empty field of vision, his and ours, within which the sentence was conceived. Even if the scene does not figure in the memoir, it says something about the relation of *The Diving Bell and the Butterfly* to the project of narrative medicine itself, mentioned in the previous chapter, advocating for the restoration of expressivity within the clinical setting. Megan Craig writes of this scene, "in his first words, Bauby reminds us that speech emerges from a body that may be wounded beyond expression."[22] Seen from that angle, *The Diving Bell and the Butterfly* registers the limits of autopathography—not in the sense that there is only so far autopathography can go as a genre, or as a quasi genre, but rather because what it registers does not always reflect back the familiar, or logical, contours of the speaking, thinking, and (life) loving subject. At the limit of autopathography, what pulls autopathography together, as the expression and continuous exercise of the subject and patient's prerogatives, falls apart, to be made available only by a script—whether as film or published memoir—to form a relational and clinical otherwise.

Bauby's memoir does not end on a return to the normative horizon of a past life and its past body image but with the acknowledgment of a new bodily state ("I have indeed begun a new life, and that life is here, in this bed, that wheelchair, and those corridors"), a new calendar ("I am savoring this last week of August with a heart that is almost light, because for the first time in a long while I don't have that awful sense of a countdown triggered at the beginning of the summer that inevitably spoils a good part of it"), and a new range of promises: "I can now grunt the little song about the kangaroo, musical testimony to my progress in speech therapy" (129–30, translation modified). In the meantime, Bauby developed a grammar of tenses grounded in a clinical environment that takes care of a

FIGURE 6.1 Screenshot of Julian Schnabel, *The Diving Bell and the Butterfly* (2007).

body that he is striving to recognize as his own: "Having refused to adopt the hideous jogging suit prescribed by the hospital, I am now dressed in the old rags I was wearing in my student days. . . . My old jackets could open painful tracks in my memory. Rather, I see in them a symbol of continuing life. And the proof that I still want to be myself" (17, translation modified). Continuity in time is not primarily indexed on the availability of a grammatically recognizable future tense, but rather on a remembering process that is also a remembering project stitching together the body before and the body after the accident. For Bauby, continuity is commemorated. Remembering is for him the tense of life after clinical survival: "By means of a tube threaded into my stomach, two or three bags of a brownish fluid provide my daily caloric needs. For pleasure, I have to turn to the vivid memory of tastes and smells, an inexhaustible reservoir of sensations" (36).

To give a point of comparison and tease out the relation between clinical survival and distinction of tenses, I turn to Dikaios Sakellariou, a specialist of occupational therapy at Cardiff University, who explains that

> when dysphagia affects quality of life and nutritional intake, percutaneous endoscopic gastrostomy (PEG) is often recommended. A PEG is an opening through the abdominal wall and into the stomach through which people can receive food, water, and medication. It is standard practice in MND and can lead to improved nutrition. Vagelio's desire was to keep her body intact, which meant no PEG, although PEG was recommended soon after diagnosis.[23]

PEG is a technology of survival ensuring the patient's daily caloric intake, but can it be for Vagelio—Dikaios's mother—a technique of existence?: "People . . . choose those technologies they can picture themselves using; they construct an imaginary future and choose those technologies that seem compatible with that anticipated and desired future. Some procedures can be associated with a dreaded rather than a desired future, especially when desire seeks more than survival." For Vagelio, continuity in time is defined by concerns pertaining to bodily integrity and issues of dignity (dignity understood as a specific form of continuity in time in medieval political theology).[24] It is not a grammatical given but a precarious field of tense differentiation and a contested site of biographical capture where physiological prospects are negotiated against ordinary affects.

"And then I sink into coma."[25] The reconstitution of the vascular accident in the penultimate chapter of *The Diving Bell and the*

Butterfly is not a locus of continuity. The stroke, in Bauby's narrative, is but an interruption in a series of interruptions, one interrupting another. The fateful day appears punctuated by other memories of other days, by lyrics of a song heard on the radio, by social obligations, by the convenience of a new car, and by the inconvenience of traffic. The missing link that is supposed to explain the passage from the body before to the body after is not a link after all. Physiology and neurology account for the accident, but the interruption that they describe will not be recovered by the patient's narrative. In other words, the stroke belongs to an unrecoverable temporality: no matter how far Bauby's account goes into the reconstitution of his loss of consciousness, no matter how exactly and accurately he manages to replace the accident in a series of events, he "creates a fictive past in the present that fills in the gaps surrounding the fall."[26] I quote from Lawrence Kritzman's insightful reading of "De l'exercitation" ("Of Practice") to suggest a parallel between Bauby's narrative and Montaigne's near-death experience following a horseback accident.

In Bauby's case, the narrative stems from a period of unconsciousness following a vascular accident. It is the narrative of an awakening to a new day ("Through the frayed curtain at my windows, a wan glow announces the break of the day") and to a new bodily state. The narrative itself is a form of awakening that is contingent upon the ongoing remaking of life and death in the age of biomedicine ("In the past, it was known as a "massive stroke," and you simply died").[27] As such, it is the memoir of an encounter between "personal contingency" and "contingency in the history of technology." Jean-Luc Nancy explains the distinction in his post–heart transplant memoir: "Had I lived earlier, I would be dead; later, I would be surviving in a different manner. But 'I' always finds itself caught in the battlements and

gaps of technical possibilities."[28] If being viable for Canguilhem, as already mentioned in the preface, is to have one shot at
life, for Montaigne, to be alive and be good at it, and to get
better at it through practice, experience, and exercise, is to have
only one shot at death: "But for dying, which is the greatest task
we have to perform, practice cannot help us . . . but as for death,
we can try it only once: we are all apprentices when we come to
it."[29] "Of Practice" is a prephysiological text in the sense that the
relationship between life and death is not articulated in the
vocabulary of physiology, but also to the extent that the text
itself gives an experimental form to the opposition between life
and death. Physiological death is a process that can be observed,
almost triangulated. Bichat, for instance, devised a protocol
involving the methodical destruction or disruption of vital
organs in animals—the brain, the lungs, and the heart—to
assess the induced effects on each other. For Montaigne, death
cannot be experienced, at least not in physiological terms, or as
a vegetative state of wavering between life and death—he is
quite clear about that: "I can imagine no state so horrible and
unbearable for me as to have my soul alive and afflicted, without
means to express itself"[30]—but it can be approximated in our
sleep: "Perhaps the faculty of sleep, which deprives us of all
action and all feeling, might seem useless and contrary to nature,
were it not that thereby Nature teaches us that she has made us
for dying and living alike, and from the start of life presents to
us the eternal state that she reserves for us after we die, to accus-
tom us to it and take away our fear of it" (268). Sleep is what
foregrounds the relationship between death and life. But sleep
is more than death by proxy, it is in its relation to waking a
trope, a lesson in flatlining, so to speak. It is the detour by
which "Montaigne tries to find a way around the impossibility
of describing the experience of death."[31] Sleep provides him

with a repertoire of sensations in which to recount the "toppling over" of consciousness and imagine what it was like to be temporarily dead and to speak as if he were already dead:

> It seemed to me that my life was hanging only by the tip of my lips; I closed my eyes in order, it seemed to me, to help push it out, and took pleasure in growing languid and letting myself go. It was an idea that was only floating on the surface of my soul, as delicate and feeble as all the rest, but in truth not only free from distress but mingled with that sweet feeling that people have who let themselves slide into sleep. I believe that this is the same state in which people find themselves who we see fainting with weakness in the agony of death; and I maintain that we pity them without cause, supposing that they are agitated by grievous pain or have their soul oppressed by painful thoughts. This has always been my view . . . concerning those whom we see thus prostrate and comatose as their end approaches, or overwhelmed by the length of the disease, or by a stroke of apoplexy, or by epilepsy . . . or wounded in the head: When we hear them groan and from time to time utter poignant sighs, or see them make certain movements of the body, we seem to see signs that they still have some consciousness left; but I have always thought, I say, that their soul and body were buried in sleep.

(269–70)

To be sure, temporary death is not the experience of death. It is an encounter in palliative time with the physiological body before physiology, a body hurt beyond feeling and beyond expression, animated but not conscious: a body that cannot be called ours and that is no longer mine. If death remains out of reach, the encounter with perishability is the next best thing, instructive enough to offset the pointlessness there would be otherwise to tell the "tale of such a trivial [*legier*] event" (272, translation modified). Thus there is a case to be made for the *case*—whether *chute*, *accident*, or *decay* (all these terms are related to the Latin verb *cadere,* to fall). At stake in this effect of biographical capture, from Montaigne to Bauby, from de Certeau to Allan King, is the invention of a form of *casual* relevance—for want of a better term—as the form of what remains at odds with the realm of policy relevance in the governance of decay.

EPILOGUE

I n Ovid's text, Myrrha's prayer finds benevolent ears in the
divinity who reinvents her body between life and death,
between burial and storage, to bear witness to a defacing
and destructive form of responsibility and accountability. In
turn, Ovid's text found a form of benevolence in the craft of two
sixteenth-century illustrators granting with images Myrrha's
impossible demand: *"Make me nothing / Human: not alive, not
dead"*—in the words of American poet Frank Bidart, who rein-
terpreted her legend in "The Second Hour of the Night" (1997).
In *La Métamorphose d'Ovide figurée* (1557), Bernard Salomon
pauses Ovid's text on the precise moment where Myrrha delivers her son Adonis. Seven nymphs in human garb, three on her
right carrying a shallow basin and an ewer, three on her left
bringing linens, assist Myrrha, while a kneeling nymph pulls
the child from the bark. In Virgil Solis's reinterpretation of the
same engraving, the newborn is hidden by the kneeling nymph's
body. Only Adonis's left hand is visible as he emerges from
the tree into the image. It is a difficult birth—for "the pain
cannot form words, nor can Lucina be called on, in the voice of
a woman in labour. Nevertheless the tree bends, like one strain-
ing, and groans constantly, and is wet with falling tears." It is

FIGURE 7.1 Johannes Posthius and Virgil Solis, *Iohan: Posthii Germershemii Tetrasticha in Ovidii Metamor* (Frankfurt, 1563), 127. Courtesy of the Bayerische Staatsbibliothek/Münchener Digitalisierungszentrum.

also a difficult birth to represent, for, according to Ovid's text, "Even Envy would praise [Adonis's] beauty, being so like one of the torsos of naked Amor painted on boards. But to stop them differing in attributes, you must add a light quiver, for him, or take theirs away from them" (vv. 515–17). It is as if, in this passage, an image had given birth to another image. Or, to put it differently, the stillbirth of Myrrha as an image of herself in the tree she became parallels the birth of a living—quivering—image of Love in Adonis.

In the same way, there is, to the best of my knowledge, no transmitted relation between the biotech and biomed present and Myrrha's story. There is no direct mention of Ovid to be found in *Camera Lucida*, Barthes's meditation on photography and the work of mourning. Both memories are derived. Both hinge on what he calls an "uncertain filiation":

> Once I feel myself observed by the lens, everything changes: I constitute myself in the process of "posing," I instantaneously make another body for myself, I metamorphose myself in advance into an image. This transformation is an active one: I feel that the Photography creates my body or mortifies it, according to its caprice. . . . No doubt it is metaphorically that I derive my existence from the photographer. But though this dependence is an imaginary one (and from the purest image-repertoire), I experience it with the anguish of an uncertain filiation: an image—my image—will be generated.[1]

As with Myrrha's metamorphosis, the birthing properties of the photographic image are part of the punishment:

> If only Photography could give me a neutral, anatomic body, a body which signifies nothing! Alas, I am condemned by (well-meaning) Photography always to have a physical expression [*une mine*]: my body never finds its zero degree, no one can give it to me (perhaps only my mother? For it is not indifference which erases the weight of the image—the Photomat always turns you into a criminal type, wanted by the police—but love, extreme love).[2]

In "The Life of Infamous Men," Foucault explained that in modern societies biography is an instrument of capture and control. It is a modality of leaving traces that occurs within specific conditions and for specific purposes. Not only does biographical capture give form to what counts and is recognized as an existence, but it also channels a discourse on the ordinary into a domain of sovereign intervention. Barthes's own metamorphosis into some body always at least potentially suspect, if not already suspected (and objectified), alerts us to the emerging situation in which perishable lives receive, as if on loan, the "words, turns of phrase and constructions, rituals of language" to speak about themselves and, sometimes, gain access to biography.[3]

A few pages later, Barthes admits: "For me the noise of Time is not sad: I love bells, clocks, watches and I recall that at first photographic implements were related to techniques of cabinet-making and the machinery of precision: cameras, in short, were clocks for seeing [*des horloges à voir*], and perhaps in me someone very old still hears in the photographic mechanism the sound of the living wood [*le bruit du bois vivant*]."[4] The last expression is quite enigmatic. Is the wood alive with the memory of the tree that it was at some point, or is it "living" by virtue of the images it engenders? Is the camera-cabinet alive with the haunting memory of Myrrha, who, too, gave birth to an image? Regardless of what one might consider to be the most viable hypothesis, the important point remains that the generation of uncertain filiations—Barthes's memory of the "living wood" or the memory of Myrrha in the age of bioengineering and biobanking—recodes a set of material-technical relations ("the machinery of precision") into a form of continuity in time ("in me someone very old still hears . . . the sound of the living wood"). This minute effect of recoding speaks to the book's larger ambitions to draw attention to what it means for protocols of conservation, prediction, and termination to "loan words, turns of phrase and constructions, rituals of language" and thus mediate our relation to loss and decay.

In a knowledge economy driven by risk management, policy relevance names something quite desirable, so desirable, in fact, that it tends to invalidate parallel modes of inquiry into the logics of containment and postponement imagined by terminal capitalism. One is left wondering, as a result, what is the good of an intervention that does not describe and assess procedures in a world where only experts can be serious about the "rules, institutions and capabilities which specify how risks are to be identified in particular contexts."[5] *Perishability Fatigue* argues that there is value in confronting and commemorating what it is like to have one's sense of continuity in time engendered by contractual practices that, in the case of biobanking, remove tissues "from the flow of historical and biological time . . . so that their potential can be realized at a future date."[6] There is value in attending to the paradox of technologies designed and deployed to contain, predict, and preserve, only to serve in the end as monuments, icons, and fables of transience. Finally, there is value in paying heed to what else is at work in the refashioning of our relation to loss, storage, and decay in the age of seed banks, GMOs, toxic waste management, tissue culture, and palliative care.

Perishable lives have not necessarily been lived. They can exist in a state of projects, projections, fables, or, conversely, as *memento mori*. They are not necessarily human. They define modes of coping with surplus and anticipating loss in a life cycle that does not seem to afford ports of entry or points of exit. In turn, the perishable lives featured throughout this book, whether they have been lived and lost, whether human or not, whether apocryphal, accidental, or authenticated, speak to something that resists policing in matters of perishability, something on the order of what Freud observed and acknowledged in a short text published in 1915. "Vergänglichkeit" ("On Transience") recounts the semipastoral anecdote of a summer walk and Freud's attempts to reason with two friends who do not seem to be able to enjoy the radiant scenery, too disturbed as they are by the thought of the perishability of everything. Freud's approach to the prospect of extinction (*dem Vergehen geweiht war*) is to bring transience into the realm of calculation and desire: "I could not see my way to dispute the transience of all things, nor could I insist upon an exception in favour of what is beautiful and perfect. But I did dispute the pessimistic poet's view that the transience of what is beautiful involves any loss in its worth. On the contrary, an increase! Transience value is scarcity value in time."[7] In this economy of attachment and detachment, perishability is a form of currency and a domain of investment, not a scene of affective collapse. The logic is solid, though it still fails to impress the two interlocutors. While the argument appears incontestable and affirming, Freud's companions in transience do not share his literacy of finitude. For them, managing perishability is not a matter of bouncing back or getting better at life by investing in "the proneness to decay" (305). For them, a culture of perishability is not the mechanism that will allow viable creatures to carry on and keep on living no matter what, or to envision continuity in time against all odds—including extinction, for, Freud ventures, "a time may indeed come when the pictures and statues which we admire today will crumble to dust, or a race of men may follow us who no longer understand the works of our poets and thinkers, or a geological epoch may even arrive when all animate life upon the earth ceases" (305). The two friends took a different path. For them, as for many others living in a meantime where there is no shortage of disaster scenarios and abundance of reasons to despair, perishability fatigue opens on a different scene, script, or scenario of what feels like loss and stings like the present.

NOTES

PREFACE: MYRRHA'S PRAYER

1. Ovid, *Metamorphoses*, 10, trans. Anthony S. Kline, vv. 431–502, http://ovid.lib.virginia.edu/trans/Metamorph10.htm#484521426.
2. Elizabeth Archibald, *Incest and the Medieval Imagination* (Oxford: Oxford University Press, 2001), 88.
3. See Sarah Lochlann Jain, *Malignant: How Cancer Becomes Us* (Berkeley: University of California Press, 2013).
4. Georges Canguilhem, "Diseases," in *Writings on Medicine*, trans. with an introduction by Stefanos Geroulanos and Todd Meyers (New York: Fordham University Press, 2012), 34–41 (41) (emphasis in the original).
5. Michel Foucault, *History of Sexuality*, vol. 1: *An Introduction*, trans. Robert Hurley (New York: Pantheon, 1980), 135 and 136.
6. Jacqueline Pfeffer Merrill, "Embryos in Limbo," *New Atlantis* 24 (Spring 2009): 18–28 (26).
7. Merrill, "Embryos in Limbo," 24. See also R. D. Nachtigall, K. Mac-Dougall, J. Harrington, J. Duff, G. Lee, M. Becker, "How Couples Who Have Undergone In Vitro Fertilization Decide What to Do with Surplus Frozen Embryos," *Fertility and Sterility* 92, no. 6 (December 2009): 2094–96.
8. Stuart Pimm, "Opinion: The Case Against Species Revival," *National Geographic*, March 12, 2013, http://news.nationalgeographic.com/news/2013/03/130312—deextinction-conservation-animals-science

-extinction-biodiversity-habitat-environment. For a brief overview of the de-extinction argument, see Steward Brandt, "The Case for De-Extinction: Why We Should Bring Back the Woolly Mammoth," *Environment 360*, January 13, 2014, http://e360.yale.edu/feature/the _case_for_de-extinction_why_we_should_bring_back_the_woolly _mammoth/2721.

9. Lauren Berlant, *Cruel Optimism* (Durham: Duke University Press, 2011).

10. Karin Knorr Cetina, "The Rise of a Culture of Life," *EMBO Reports* 6 (2005): 76–80 (76).

11. Karen Pinkus and Cameron Tonkinwise, "Want Not: A Dialogue on Sustainability with Images," *World Picture* 5 (Spring 2011): 1–11 (6), http://www.worldpicturejournal.com/WP_5/PDFs/Pinkus_Tonkin wise.pdf.

12. Cameron Tonkinwise, "Practicing Sustainability by Design: Global Warming Politics in a Post-awareness World," *Scapes* 6 (Fall 2007): 4–12.

13. Cameron Tonkinwise, "Sustainability Is Not a Humanism: Review Essay on Allan Stoekl's *Bataille's Peak*," *Design Philosophy Papers* 7, no. 1 (2009): 39–48 (41–42).

14. Gillen D'Arcy Wood, "What Is Sustainability?" *American Literary History* 24, no. 1 (2012): 1–15 (10).

15. United Nations, *Our Common Future: The World Commission on Environment and Development* (Oxford: Oxford University Press, 1987), 40.

16. See Michel Serres, *Malfeasance: Appropriation Through Pollution?*, trans. Anne-Marie Feenberg-Dibon (Stanford: Stanford University Press, 2010).

17. Donald Winnicott, *Playing and Reality* (London: Routledge, 1991), 17.

18. Jason Pine, "Economy of Speed: The New Narco-Capitalism," *Public Culture* 19, no. 2 (2007): 357–66.

19. Revisiting the case, Elizabeth Povinelli suggests that in trying to "undo the damage done by the arrival of his sister . . . the boy created new habitations, new ways of being held. He did not mean to do this, but his refusal was a creative act. It provided an environment for alternative possibilities of life." Elizabeth A. Povinelli, "After the Last Man: Images and Ethics of Becoming Otherwise," *e-flux* 35

(May 2012), http://www.e-flux.com/journal/35/68380/after-the-last-man
-images-and-ethics-of-becoming-otherwise/.

20. Joseph Masco, *The Nuclear Borderlands: The Manhattan Project in Post–
Cold War New Mexico* (Princeton: Princeton University Press, 2006),
27.

21. See Hermann Kahn, *On Thermonuclear War* (Princeton: Princeton
University Press, 1960).

1. BEING FABULOUS AS THE CLIMATE CHANGES

1. Norwegian Ministry of Agriculture and Food website, http://www
.regjeringen.no/en/dep/lmd/campain/svalbard-global-seed-vault
/description.html?id=464076.

2. Kenny Ausubel, *Seeds of Change: The Living Treasure* (San Francisco:
Harper Collins, 1994), 20. On biodiversity as cultural value, see Sarah
Franklin, Celia Lury, and Jackie Stacey, *Global Nature, Global Culture*
(London: Sage, 2000), 79.

3. http://www.heatherwick.com/uk-pavilion/.

4. Lars Peder Brekk, "Frozen Seeds in a Frozen Mountain—Feeding a
Warming World!" February 26, 2009, http://www.regjeringen.no/en
/dep/lmd/whats-new/Speeches-and-articles/speeches-and-articles
-by-the-minister/speeches-and-articles-/one-year-anniversary
-seminar-of-the-sval/one-year-anniversary-seminar-of-the-sval
.html?id=547254.

5. Jacques Derrida, "Of an Apocalyptic Tone Newly Adopted in Philosophy," in *Derrida and Negative Theology*, ed. Harold G. Coward and
Toby Foshay (Albany: SUNY Press, 1992), 25–71 (64).

6. See Karen Pinkus, "Carbon Management: A Gift of Time?" *Oxford
Literary Review* 32, no. 1 (2010): 51–70.

7. In an interview with Cary Fowler, the executive director of the Global
Crop Diversity Trust, Ross Andersen says, "when I think about the
seed vault, the first thing that stands out to me is that it's really
a technology of deep time, a way of coping with the kinds of events
that happen on very broad time scales." Ross Andersen, "After Four
Years, Checking Up on the Svalbard Global Seed Vault," *Atlantic*,
February 28, 2012. Politics is always a politics of the future; in Lee

Edelman's words: "we are no more able to conceive of a politics without a fantasy of the future than we are able to conceive of a future
without the figure of the Child." Lee Edelman, *No Future: Queer
Theory and the Death Drive* (Durham: Duke University Press, 2005), 11.
For a pre-Edelman take on reproductive futurism, see Frances Ferguson, "The Nuclear Sublime," *Diacritics* 14, no. 2 (Summer, 1984): 4–10.
Puzzlingly enough, or perhaps not, the reference to Lee Edelman is
almost always absent from the discourse of sustainability studies, with
the exception being Karen Pinkus, "The Risks of Sustainability," in
*Criticism, Crisis, and Contemporary Narrative. Textual Horizons in an
Age of Global Risk*, ed. Paul Crosthwaite (London: Routledge, 2011),
62–80.

8. Allan Stoekl, "'After the Sublime,' After the Apocalypse: Two Versions of Sustainability in Light of Climate Change," *diacritics* 41, no. 3
   (2013): 40–57 (41).

9. François Rabelais, *The Fourth Book*, in *The Complete Works of François
   Rabelais*, trans. Donald M. Frame (Berkeley: University of California
   Press, 1991), 556. François Rabelais, *Œuvres complètes*, ed. Mireille
   Huchon, with François Moreau (Paris: Gallimard-NRF, 1994), 667:
   "Compaignons, oyez-vous rien? Me semble que je oy quelques gens
   parlans en l'air, je n'y voy toutesfoys personne. Escoutez."

10. Rabelais, *The Fourth Book*, 557. Rabelais, *Œuvres complètes*, 668–69:

    J'ai leu qu'un Philosophe, nommé Petron, estoyt en ceste opinion
    que feussent plusieurs mondes soy touchans les uns les aultres en
    figure triangulaire aequilaterale, en la pate et centre des quelz disoit
    estre le manoir de Vérité, et là habiter les Parolles, les Idées, les
    Exemplaires et portraicts de toutes choses passées et futures: autour
    d'icelles estre le Siecle. Et en certaines années, par longs intervalles,
    part d'icelles tomber sus les humains comme catarrhes, et comme
    tomba la rousée sus la toizon de Gedeon: part là rester reservée pour
    l'advenir, jusques à la consommation du Siecle. Me souvient aussi
    que Aristoteles maintient les parolles de Homere estre voltigeantes,
    volantes, moventes, et par consequent animées. D'adventaige Anti-
    phanes disoit la doctrine de Platon es parolles estre semblable
    lesquelles en quelque contrée en temps du fort hyver lors que sont
    proferées, gelent et glassent à la froydeur de l'air, et ne sont ouyes.

11. Rabelais, *Œuvres complètes*, 670.

12. Rabelais, *The Fourth Book*, 559. Rabelais, *Œuvres complètes*, 670:

> hin, hin, hin, hin, his, ticque, torche, lorgne, brededin, brededac, frr, frrr, frrr, bou, bou, bou, bou, bou, bou, bou, traccc, trac, trr, trr, trr, trrr, trrrr. On, on, on, on, ouououon: goth, magoth, et ne sçay quelz aultres motz barbares, et disoyt que c'estoient vocable du hourt et hannissement des chevaulx à l'heure qu'on chocque, puys en ouysme dèaultres grosses et rendoient son en degelent, les unes comme de tabours, et fifres, les aultres comme de clerons et trompettes.

13. Marion Leathers Kuntz, "Rabelais, Postel et utopie," *Travaux d'Humanisme et Renaissance*, no. 321 *Études Rabelaisiennes*, vol. 33, *Rabelais pour le XXIe siècle: Actes du colloque du Centre d'Etudes Supérieures de la Renaissance (Chinon-Tours, 1994)*, ed. Michel Simonin (Geneva: Droz, 1998), 55–78 (61).

14. Rabelais, *The Fourth Book*, 559 (translation modified). Rabelais, *Œuvres complètes*, 670–71: "Je vouloys quelques motz de gueule mettre en reserve dedans de l'huille, comme l'on guarde la neige et la glace, et entre du feurre bien nect. Mais Pantagruel ne le voulut, disant estre follie faire reserve de ce dont jamais l'on n'a faulte et que tous jours on a en main, comme sont motz de gueule entre tous bons et joyeulx Pantagruelistes." On the metadiscursive dimension of this passage, see Michel Jeanneret, "Les paroles dégelées (Rabelais, *Quart Livre*, 48–65)," *Littérature* 17 (1975): 14–30.

15. Friedrich Kittler, "Gramophone, Film, Typewriter," trans. Dorothea von Mücke and Philippe L. Similon, *October* 41 (Summer 1987): 101–18 (104).

16. Karen Pinkus, *Fuel: A Speculative Dictionary* (Minneapolis: University of Minnesota Press, 2016).

17. Matthew Liao, Anders Sandberg, and Rebecca Roache, "Human Engineering and Climate Change," *Ethics, Policy, and Environment* 15, no. 2 (June 2012): 206–21 (216–17).

18. I borrow the term *panoply* from Michel de Certeau, *The Practice of Everyday Life*, trans. Steven Rendall, 2 vols. (Berkeley: University of California Press, 1984), 1:142.

19. Jean-François Lyotard, "*Oikos*," in *Political Writings*, trans. Bill Readings and Kevin Geiman (Minneapolis: University of Minnesota Press, 1993): 96–107 (106).

20. Jean-François Lyotard, *Postmodern Fables*, trans. Georges Van Den Abbeele (Minneapolis: University of Minnesota Press, 1997), 91. On the relation between reality and realism, Lyotard writes: "Realism is the art of making reality, of knowing reality and knowing how to make reality. The story we just heard says that this art will still develop a lot in the future. . . . The fable is realist because it recounts the story that makes, unmakes, and remakes reality."

21. For another example of this conflation, see Andy Miah, "Justifying Human Enhancement: The Accumulation of Biocultural Capital," in *The Transhumanist Reader: Classical and Contemporary Essays on the Science, Technology, and Philosophy of the Human Future*, ed. M. More and N. Vita-More (Malden, MA: Wiley-Blackwell, 2013), 291–301 (297): "the strongest value claim that one can make in relation to body modifications is conceptually no different from the value claim one might make about reading a book or watching a movie."

22. Jean-François Lyotard, "Prescription," trans. Christopher Fynsk, *L'Esprit Créateur* 31, no. 1 (1991): 15–32 (24).

23. Ovid, *Metamorphoses*, 10, vv. 499–500.

24. Franz Kafka, "In the Penal Colony," trans. Ian Johnston, *The Kafka Project*: "'You are free now,' said the Officer. For the first time the face of the Condemned Man showed signs of real life." ["'Nun, frei bist du', sagte der Offizier. Zum erstenmal bekam das Gesicht des Verurteilten wirkliches Leben."], http://www.kafka.org/index.php?aid=167.

2. STILL LIFE WITH GENETICALLY MODIFIED TOMATO

1. *Biotechnology Law Report* 8, no. 6 (November-December 1989), 489–90 (489). See also G. Bruening and J. M. Lyons, "The Case of the FLAVR SAVR Tomato," *California Agriculture* 54, no. 4 (July-August 2000): 6–7; and Paul Rabinow, "Artificiality and Enlightenment: From Sociobiology to Biosociality," in *Essays on the Anthropology of Reason* (Princeton: Princeton University Press, 1996), 91–111 (105–7).

2. Suman Sahai, "The Bogus Debate of Bioethics," *Biotechnology and Development Monitor* 30 (March 1997), 24.
3. Sahai, "The Bogus Debate of Bioethics."
4. Vandana Shiva, "Bioethics: A Third World Issue," http://www.nativeweb.org/pages/legal/shiva.html.
5. See Anne-Lise François, "Flower Fisting," *Postmodern Culture* 22, no. 1 (2011) and "'O Happy Living Things' Frankenfoods and the Bounds of Wordsworthian Natural Piety," *diacritics* 33, no. 2 (2005): 42–70.
6. Dan Glickman cited in Anne-Lise François, "'O Happy Living Things' Frankenfoods," 44.
7. Norman Bryson, *Looking at the Overlooked: Four Essays on Still Life Painting* (Cambridge: Harvard University Press, 1990), 13–14.
8. In *Still Life and Trade in the Dutch Golden Age* (New Haven: Yale University Press, 2007), art historian Julie Berger Hochstrasser proposed to recover the commercial energy and the imperial rapacity that fueled Dutch still life painting.
9. Bryson, *Looking at the Overlooked*, 132.
10. This descent into what Harry Berger calls "small scale violence" complicates the iconographic interpretation of still life painting as a means to preserve the visual intensity of the "caterpillage" against allegorical reductions to a *vanitas* motif. Harry Berger, *Caterpillage: Reflections on Seventeenth-Century Dutch Still Life Painting* (New York: Fordham University Press, 2011).
11. Bryson, *Looking at the Overlooked*, 121–22.
12. Bryson, *Looking at the Overlooked*, 120.
13. Adam Smith, *An Inquiry Into the Nature and Causes of the Wealth of Nations* (1776), II, 3, §1.
14. Ernst H. Gombrich, "Tradition and Expression in Western Still-life," *Burlington Magazine*, no. 103 (1961): 175–80 (180), quoted by Hochstrasser, *Still Life and Trade in the Dutch Golden Age*, 257.
15. Hannah Landecker, "Food as Exposure: Nutritional Epigenetics and the New Metabolism," *BioSocieties* 6 (2011): 167–94 (177).
16. See Hanneke Grootenboer, "Perspective as Allegorical Form: *Vanitas* Painting and Benjamin's Allegory of Truth," in *The Rhetoric of Perspective: Realism and Illusionism in Seventeenth- Century Dutch Still-Life Painting* (Chicago: University of Chicago Press, 2005), 135–65.

17. Bryson, *Looking at the Overlooked*, 104.
18. Charles Sterling, *Still Life Painting: From Antiquity to the Twentieth Century* (New York: Harper and Row, 1981), 158 (my emphasis).
19. Bryson, *Looking at the Overlooked*, 128.
20. Norwegian Ministry of Agriculture and Food, "Svalbard Global Seed Vault," http://www.regjeringen.no/en/dep/lmd/campain/sval bard-global-seed-vault.html?id=462220. For a dramatic illustration of this phenomenon of disjuncture, see John Seabrook, recounting the sacrifice of the Varilov Research Institute of Plant Industry members in St. Petersburg in 1941–42 who starved to death in the middle of the seed archive they had promised to protect. "Sowing for Apocalypse: The Quest for a Global Seed Bank," *New Yorker*, August 27, 2007.
21. Lauren Berlant, "Slow Death (Obesity, Sovereignty, Lateral Agency)," in Lauren Berlant, *Cruel Optimism* (Durham: Duke University Press, 2011), 95–119.
22. See Paolo Palladino, *Plants, Patients, and the Historian: (Re)membering in the Age of Genetic Engineering* (Manchester: Manchester University Press, 2002).

3. STORE AND TELL

1. Karl Marx, *The Capital*, I, VIII, 26.
2. See, for instance, Myra Hird, "Waste, Landfills, and an Environmental Ethic of Vulnerability," *Ethics and the Environment* 18, no. 1 (2013): 105–24.
3. Peggy Kamuf, "Life in Storage: Of Capitalism and A&E's 'Storage Wars,'" *Los Angeles Review of Books*, February 4, 2012, https://lareviewofbooks.org/article/life-in-storage-of-capitalism-and-aes-storage-wars/#!
4. In the same manner, Elissa Marder notes that "the 'womb' that holds the body before the beginning of life is structurally indistinguishable from the 'tomb' that holds the body after the end of life." Elissa Marder, *The Mother in the Age of Mechanical Reproduction: Psychoanalysis, Photography, Deconstruction* (New York: Fordham University Press, 2012), 2–3.
5. Julia Leigh, *The Hunter* (New York: Four Walls Eight Windows, 1999).

6. Alexander Klose, *The Container Principle: How a Box Changes the Way We Think,* trans. Charles Marcrum (Cambridge: MIT Press, 2015), 152.

7. Klose, *The Container Principle,* 317.

8. Jean-Pierre Vernant, "At Man's Table: Hesiod's Funcatin Myth of Sacrifice," in *The Cuisine of Sacrifice Among the Greeks,* ed. Marcel Detienne and Jean-Pierre Vernant, trans. Paula Wissing (Chicago: Chicago University Press, 1989), 21–86 (79).

9. Michel de Certeau, *The Writing of History,* trans. Tom Conley (New York: Columbia University Press, 1988), 100; Leerom Medovoi, "A Contribution to the Critique of Political Ecology: Sustainability as Disavowal," *New Formations* 69 (2010): 129–43 (142).

10. For a reconstitution of the logic of the primal murder, see Philip Lewis, *Seeing Through the Mother Goose Tales: Visual Turns in the Writings of Charles Perrault* (Stanford: Stanford University Press, 1996), 208.

11. On the question of transmissibility in folktales, see Peter Brooks, *Reading for the Plot: Design and Intention in Narrative* (Cambridge: Harvard University Press, 1992), 27.

12. Charles Perrault, "The Blue Beard" cited in Winfried Menninghaus, *In Praise of Nonsense: Kant and Bluebeard,* trans. Henry Pickford (Stanford: Stanford University Press, 1999), 58 (translation modified).

13. Charles Perrault, *Histoires ou Contes du temps passé: Avec des moralitez* (Paris: Barbin, 1697), 81.

14. Menninghaus, *In Praise of Nonsense,* 69–70: "The *moralité* becomes a *moralité* about the superfluity of a recit which no longer provides any useful lesson."

15. Thomas A. Sebeok, *Communication Measures to Bridge Ten Millennia* (Columbus, OH: Office of Nuclear Waste Isolation, Battelle Memorial Institute, 1984), iv.

16. Toni Feder, "DOE Opens WIPP for Nuclear Waste Burial," *Physics Today* 52, no. 5 (1999): 59.

17. Peter van Wyck, *Signs of Danger: Waste, Trauma, and Nuclear Threat* (Minneapolis: University of Minnesota Press, 2005), 88.

18. Georgius Agricola, *De Re Metallica,* trans. Herbert Clark Hoover and Lou Henry Hoover (New York, Dover, 1950), 218.

19. Sebeok, *Communication Measures to Bridge Ten Millennia*, 28.
20. Ross Andersen, "After Four Years, Checking Up on the Svalbard Global Seed Vault," *Atlantic*, February 28, 2012. On participatory preparedness, see also Elaine Scarry, *Thinking in an Emergency* (New York: Norton, 2011). On the notion of "long-term stewardship," see Joseph Masco, "Mutant Ecologies: Radioactive Life in Post–Cold War New Mexico," *Cultural Anthropology* 19, no. 4 (2004): 517–50 (535).
21. Walter Benjamin, "The Storyteller: Reflections on the Works of Nikolai Leskov," in *Illuminations*, ed. Hannah Arendt, trans. Harry Zohn (New York: Schocken, 2007), 83–110 (90).
22. "How Will Future Generations Be Warned?" http://www.wipp.energy.gov/fctshts/PICs.pdf.
23. Myra Hird, "Landscapes of Terminal Capitalism, Aporias of Responsibility: Lifeworlds Inherited, Inhabited, and Bequeathed," keynote address, "Anthropocene Feminism," conference at the Center for Twenty-First-Century Studies, University of Wisconsin-Milwaukee, April 2014. By derivation, terminal humanities take the measure of how, and to what degree, resource shortage, conservation politics, and concurrent investments in biocapital have altered and continue to alter the sense of a debt to life that was paid off by giving oneself to a discipline, by being productive, and reproducing oneself academically while—at least according to a recent study—mortgaging one's own fertility. See C. Lampic, A. Skoog Svanber, P. Karlström, and T. Tydén, "Fertility Awareness, Intentions Concerning Childbearing, and Attitudes Towards Parenthood Among Female and Male Academics," *Human Reproduction* 21, no. 2 (February 2006), 558–64; and Robin Wilson, "Timing Is Everything: Academe's Annual Baby Boom," *Chronicle of Higher Education*, June 25, 1999, http://chronicle.com/article/Timing-Is-Everything-/2635/.
24. Lewis, *Seeing Through the Mother Goose Tales*, 208.
25. Hans Ulrich Gumbrecht, "Shall We Continue to Write Histories of Literature?", *New Literary History*, 39 no. 3 (2008): 519–32 (519).
26. Richard G. Mitchell, *Dancing at Armageddon: Survivalism and Chaos in Modern Times* (Chicago: Chicago University Press, 2002), 5.
27. Bill Readings, *The University in Ruins* (Cambridge: Harvard University Press, 1997), 89.

28. Mitchell, *Dancing at Armageddon*, 13.
29. Mitchell, *Dancing at Armageddon*, 228.
30. Joseph Masco, "Life Underground: Building the Bunker Society," *Anthropology Now* 1, no. 2 (September 2009): 13–29 (13).
31. Jay Swayze created shelter models with trompe l'oeil windows and even one with subterranean backyard. Swayze's designed Girard B. Henderson shelter in Nevada contains a temperature-controlled weather-mimicking landscaped outdoor area with murals illustrating familiar places from the owner's childhood. Susan Roy, *Bomboozled: How the U.S. Government Misled Itself and Its People Into Believing They Could Survive a Nuclear Attack* (New York: Pointed Leaf, 2011), 120.
32. Joseph Masheck, "Alberti's 'Window': Art-Historiographic Notes on an Antimodernist Misprision," *Art Journal* 50, no. 1 (Spring 1991): 34–41.
33. Leon Battista Alberti, *On Painting and on Sculpture: The Latin Texts of De Pictura and De Statua*, ed. and trans. Cecil Grayson (London: Phaidon, 1972), 55.
34. Masco, "Life Underground: Building the Bunker Society," 24.
35. Agricola, *De Re Metallica*, 216.
36. Bill Readings, "University Without Culture," *New Literary History* 26, no. 3 (1995): 465–92 (489).
37. Readings, *The University in Ruins*, 151.
38. On philanthropy and humanistic inquiry in American institutions, see Geoffrey Galt Harpham, "The Humanities in America," keynote address at conference, "The Humanities Into the Twenty-First Century," University of Copenhagen.
39. Readings, "University Without Culture," 488.
40. William Paulson, *Literary Culture in a World Transformed: A Future for the Humanities* (Ithaca: Cornell University Press, 2001), 1. The same liminal question ("Do the humanities have a future?") appears in Samuel Weber's chapter "The Future of the Humanities: Experimenting," in *Institution and Interpretation*, expanded ed. (Stanford: Stanford University Press, 2001), 236. Geoffrey Galt Harpham, "The Humanities' Value," *Chronicle of Higher Education*, March 20, 2009, B6.

41. See Scott Herring, *The Hoarders: Material Deviance in Modern American Culture* (Chicago: University of Chicago Press, 2014). On the geopolitical underpinnings and ramifications of the belief in the availability of dumping zones, see Rob Nixon, *Slow Violence and the Environmentalism of the Poor* (Cambridge: Harvard University Press, 2011), 1–3.

42. See Allan Stoekl, *Bataille's Peak: Energy, Religion, and Postsustainability* (Minneapolis: University of Minnesota Press, 2007), 145–49; and "Agnès Varda and the Limits of Gleaning," *world picture* 5 (Spring 2011), http://www.worldpicturejournal.com/WP_5/Stoekl.html; and "Gift, Design and Gleaning," *Design Philosophy Papers* 7, no. 1 (2009): 7–17.

43. On embryo disposal, see also Mette Svendsen, "Articulating Potentiality: Notes on the Delineation of the Blank Figure in Human Embryonic Stem Cell Research," *Cultural Anthropology* 26, no. 3 (2011): 414–437; and Lynn Morgan, "'Properly Disposed Of': A History of Embryo Disposal and the Changing of Claims on Fetal Remains," *Medical Anthropology* 21, nos. 3–4 (2002): 247–74.

4. THE MORTAL LIFE OF HELA

1. Arjun Appadurai, "Grassroots Globalization and the Research Imagination," *Public Culture* 12, no. 1 (2000): 1–19 (12).

2. Tod Chambers, *The Fiction of Bioethics: Cases as Literary Texts* (New York: Routledge, 1999), 81–96.

3. Marlon Rachquel Moore, "Opposed to the Being of Henrietta: Bioslavery, Pop Culture, and the Third Life of HeLa cells," *Medical Humanities* 43 (2017): 55–61 (57).

4. Hortense J. Spillers, "Mama's Baby, Papa's Maybe: An American Grammar Book," *Diacritics* 17, no. 2 (1987): 64–81 (67) (emphasis in the original). On the link between legal issues of ownership in biotechnology and the memory of slavery in the U.S., see Priscilla Wald, "What's in a Cell?: John Moore's Spleen and the Language of Bioslavery," *New Literary History* 36, no. 2 (2005): 205–25.

5. Spillers, "Mama's Baby, Papa's Maybe," 67.

6. Hannah Landecker, "Immortality, In Vitro: A History of HeLa Cell Line," in *Biotechnology and Culture: Bodies, Anxieties, Ethics*, ed. Paul Brodwin (Bloomington: Indiana University Press, 2000), 53–72.

7. Alexander G. Weheliye, *Habeas Viscus: Racializing Assemblages, Biopolitics, and Black Feminist Theories of the Human* (Durham: Duke University Press, 2014), 79.

8. Marcel Mauss, "A Category of the Human Mind: The Notion of Person; The Notion of Self," in *The Category of the Person: Anthropology, Philosophy, History*, ed. Michael Carrithers, Steven Collins, and Steven Lukes, trans W. D. Halls (Cambridge: Cambridge University Press, 1985), 1–25 (17).

9. *Assemblage* is the critical term used by Weheliye in *Habeas Viscus* to describe the process of racialization. I use the term by reference to the volume edited by Aihwa Ong and Stephen J. Collier: "An assemblage is the product of multiple determinations that are not reducible to a single logic. The temporality of an assemblage is emergent. It does not always involve new forms, but forms that are shifting, in formation, or at stake." *Global Assemblages: Technology, Politics, and Ethics as Anthropological Problems* (Malden, MA: Blackwell, 2005), 12.

10. See Andrea Carlino, "The Book, the Body, the Scalpel: Six Engraved Title Pages for Anatomical Treatises of the First Half of the Sixteenth Century," *RES: Anthropology and Aesthetics* 16 (1988): 33–50.

11. Christine Montross, *Body of Work: Meditations on Mortality from the Human Anatomy Lab* (New York: Penguin, 2008), 20.

12. Rebecca Skloot, *The Immortal Life of Henrietta Lacks* (New York: Crown, 2010), 235.

13. Hans Belting, *An Anthropology of Images: Picture, Medium, Body*, trans. Thomas Dunlap (Princeton: Princeton University Press, 2011), 83.

14. Paul Rabinow, *Essays on the Anthropology of Reason* (Princeton: Princeton University Press, 1996), 150.

15. For an in-depth analysis of the legal dimensions of Moore's case, see James Boyle, *Shamans, Software, and Spleens: Law and the Construction of the Information Society* (Cambridge: Harvard University Press, 1997). See also the chapter that Skloot devotes to Mo cells in *The Immortal Life of Henrietta Lacks*, 199–206.

16. Paul Rabinow, *Anthropos Today: Reflections on Modern Equipment* (Princeton: Princeton University Press, 2003), 14.

17. Gilles Deleuze, *Foucault* (London: Continuum, 1999), 102.

18. Rabinow, *Essays on the Anthropology of Reason*, 149.

19. Quoted in Rabinow, *Essays on the Anthropology of Reason*, 145.

20. Ernst Kantorowicz, *The King's Two Bodies: A Study in Medieval Political Theology* (Princeton: Princeton University Press, 1957); and Caroline Walker Bynum, "Material Continuity, Personal Survival, and the Resurrection of the Body: A Scholastic Discussion in Its Medieval and Modern Contexts," *Fragmentation and Redemption: Essays on Gender and the Human Body in Medieval Religion* (New York: Zone, 1991), 239–97.

21. Rabinow, *Essays on the Anthropology of Reason*, 146–48.

22. "The Visible Human Project." U.S. National Library of Medicine, http://www.nlm.nih.gov/research/visible/visible_human.html.

23. Michel Foucault, "The Life of Infamous Men," in *Michel Foucault: Power, Truth, Strategy*, ed. Meaghan Morris and Paul Patton (Sydney: Feral, 1979), 76–91 (79).

24. Foucault, "The Life of Infamous Men," 81.

25. Giorgio Agamben, *Remnants of Auschwitz: The Witness and the Archive*, trans. Daniel Heller-Roazen (Brooklyn: Zone, 2002), 143.

26. Catherine Waldby, *The Visible Human Project: Informatic Bodies and Posthuman Medicine* (London: Routledge, 2000), 14.

27. Waldby, *The Visible Human Project*, 57.

28. Rebecca Skloot, "The Immortal Life of Henrietta Lacks, the Sequel," *New York Times*, March 23, 2013.

29. Lisa Cartwright, "The Visible Man: The Man Criminal Subject as Biomedical Norm," in *Processed Lives: Gender and Technology in Everyday Life*, ed. Melodie Calvert and Jennifer Terry (New York: Routledge, 1997), 86–100 (95–96). For a sophisticated account of the multiple inflections of the principle of privacy in the U.S., see Karla F. C. Holloway, *Private Bodies, Public Texts: Race, Gender, and Cultural Bioethics* (Durham: Duke University Press, 2011).

30. Waldby, *The Visible Human Project*, 51.

31. On the notion of transparency, see Jose van Dijck, *The Transparent Body: A Cultural Analysis of Medical Imaging* (Seattle: University of Washington Press, 2005).

32. Waldby, *The Visible Human Project*, 74.

33. http://www.frankschott.com/project/1231/process.html.

34. Philip Brophy, "Horrality—the Textuality of Contemporary Horror Films," in *The Horror Reader*, ed. Ken Gelder (London: Routledge, 2000), 276–84 (280).

35. If body horror is the other of anatomy, it also exists in opposition to the resurrection narrative. Body horror postpones indefinitely the moment when, in the resurrection narrative, the corpse goes missing, and that the evangelical trickster registering this absence delivers his message to those who came looking for it: "He is not here; for he has been raised." Matthew 28:6, quoted in Louis Marin, *On Representation*, trans. Catherine Porter (Stanford: Stanford University Press, 2002), 121.

36. Spillers, "Mama's Baby, Papa's Maybe," 67, 68 (emphasis in original).

37. See Hannah Landecker, *Culturing Life: How Cells Became Technologies* (Cambridge: Harvard University Press, 2010), 232.

38. Catherine Waldby, "Stem Cells, Tissue Cultures and the Production of Biovalue," *Health: An Interdisciplinary Journal for the Social Study of Health, Illness, and Medicine* 6, no. 3 (2002): 305–23.

39. I borrow some of the wording from Elizabeth Povinelli's lecture at ICI Berlin, "New Media/Other Worlds?" (October 2011), http://www.ici-berlin.org/de/docu/povinelli/.

40. Roland Barthes, "The World as Object," in *Critical Essays* (Evanston: Northwestern University Press, 2000), 3–12 (9–10).

41. Skloot, *The Immortal Life of* Henrietta Lacks, 1.

42. Elissa Marder, *The Mother in the Age of Mechanical Reproduction: Psychoanalysis, Photography, Deconstruction* (New York: Fordham University Press, 2012), 138–39.

5. ONCOSCRIPTS

1. See, for instance, Thomas G. Couser, *Signifying Bodies: Disability in Contemporary Life Writing* (Ann Arbor: University of Michigan Press, 2009).

2. Michel Foucault, "The Lives of Infamous Men," in *Michel Foucault: Power, Truth, Strategy*, ed. Meaghan Morris and Paul Patton (Sydney: Feral, 1979), 76–91 (89).

3. Arlette Farge, "Michel Foucault et les archives de l'exclusion ('La vie des hommes infâmes')," in *Penser la folie: Essais sur Michel Foucault*, ed. Elisabeth Roudinesco (Paris: Galilée, 1992), 65–78 (77).

4. See Lynne Huffer, *Mad for Foucault: Rethinking the Foundations of Queer Theory* (New York: Columbia University Press, 2009), 248.

5. Foucault, "The Life of Infamous Men," 91.

6. Alison Kafer, *Feminist, Queer, Crip* (Bloomington: Indiana University Press, 2013), 1.

7. Foucault, "The Life of Infamous Men," 84.

8. Nicholas A. Christakis, *Death Foretold: Prophecy and Prognosis in Medical Care* (Chicago: University of Chicago Press, 1999), 135.

9. Sarah Lochlann Jain, "Living in Prognosis: Toward an Elegiac Politics," *Representations* 98 (Spring 2007): 77–92 (90).

10. Lochlann Jain, "Living in Prognosis," 79.

11. Sarah Lochlann Jain, *Malignant: How Cancer Becomes Us* (Berkeley: University of California Press, 2013), 60. On fertility treatment within the context of cancer culture, see 128–50.

12. Rita Charon, *Narrative Medicine: Honoring the Stories of Illness* (New York: Oxford University Press, 2008), 212.

13. Georges Canguilhem, *The Normal and the Pathological,* trans. Carolyn R. Fawcett, in collaboration with Robert S. Cohen (New York: Zone, 1989), 92.

14. Todd Meyers, *The Clinic and Elsewhere: Addiction, Adolescents, and the Afterlife of Therapy* (Seattle: University of Washington Press, 2013), 10–11.

15. See, for instance, Ann Jurecic, *Illness as Narrative* (Pittsburgh: University of Pittsburgh Press, 2012).

16. Lisa Diedrich, *Treatments: Language, Politics, and the Culture of Illness* (Minneapolis: University of Minnesota Press, 2007).

17. Canguilhem, *The Normal and the Pathological,* 93.

18. Catherine Belling, "Narrating Oncogenesis: The Problem of Telling When Cancer Begins," *Narrative* 18, no. 2 (2010): 229–47 (241).

19. Barbara Ehrenreich, "Welcome to Cancerland: A Mammogram Leads to a Cult of Pink Kitsch," *Harper's* (November 2001): 43–53 (44).

20. I borrow from Michel Foucault's prose in *The Hermeneutics of the Subject: Lectures at the Collège de France, 1981–1982* (New York: Picador, 2005), 19: "such as he is, the subject is capable of truth, but that, such as it is, the truth cannot save the subject."

21. Kathleen Woodward, "Statistical Panic," *differences: A Journal of Feminist Cultural Studies* 11, no. 2 (1999): 177–203. The reference to this particular episode of *Chicago Hope* also appears in Ilana Löwy, *Preventive Strikes: Women, Precancer, and Prophylactic Surgery* (Baltimore: Johns Hopkins University Press, 2010), 199.

22. Löwy, *Preventive Strikes,* 6.

23. Barbara Duden, "Die Verkrebsung: Kränkende Diagnostik durch Krebsprävention," in *Die Gene im Kopf – der Fötus im Bauch. Historiches zum Frauenkörper* (Hanover: Offizin), 166–86 (166): "The historian knows that notions of 'cancer' and 'risk' were foreign to the past."

24. Barbara Duden, "A Historian's 'Biology': On the Traces of the Body in a Technogenic World," *Historein* 3 (2001): 89–102 (101); "Die Verkrebsung," 186: "Gerade in jenen Falten und Buchten des Leibes, die traditionell als die Orte intensiver, lebensspendender Lebendigkeit verstanden wurden, als Orte von Wunsch, Lust und Sehnsucht."

25. Lochlann Jain, *Malignant*, 29.

26. Lochlann Jain, *Malignant*, 29.

27. Sandy Flitterman-Lewis, "From *Déesse* to *Idée*: *Cleo from 5 to 7*," in *To Desire Differently: Feminism and the French Cinema*. Revised Edition (New York: Columbia University Press, 1996), 268–84 (268). On the motif of personal transformation in contemporary cancer culture, see Ehrenreich, "Welcome to Cancerland," 50–51.

28. On "landscapes of risk," see Stacy Alaimo, "Material Memoirs: Science, Autobiography, and the Substantial Self," in *Bodily Natures: Science, Environment, and the Material Self* (Bloomington: Indiana University Press, 2010), 85–112.

29. Jennifer Aniston, *Five* (US, 2011).

30. Kathleen Stewart, *Ordinary Affects* (Durham: Duke University Press, 2007).

31. Daniel Arasse, "Le corps fictif de Sébastien et le coup d'œil d'Antonello," in *Le corps et ses fictions*, ed. Claude Reichler (Geneva: Droz, 1983), 55–72.

32. Barbara Duden, *Disembodying Women: Perspectives on Pregnancy and the Unborn*. Trans. Lee Hoinacki (Cambridge: Harvard University Press, 1993), 86.

33. Duden, *Disembodying Women*, 91.

34. M. K. Bryson and Chase Joynt, "Under The Skin: Imag(in)ing Medicine's Queer Pedagogies as Moving Pictures," *No More Potlucks*, 29 (2013), http://nomorepotlucks.org/site/under-the-skin-imagining-medi cinesqueer-pedagogies-as-moving-pictures-chase-joynt-m-k-bryson.

35. Georges Didi-Huberman, "Ex-Voto: Image, Organ, Time," *L'Esprit Créateur* 47, no. 3 (2007): 7–16 (13).

36. Didi-Huberman, "Ex-Voto," 9 (emphasis in original).

37. Donna Haraway, *Modest_Witness@Second_Millennium.FemaleMan _Meets_OncoMouse: Feminism and Technoscience* (New York: Rout-ledge, 1997), xiv.

38. Haraway, *Modest_Witness@Second_Millennium*.

39. Didi-Huberman, "Ex-Voto: Image, Organ, Time," 8 (emphasis in original).

40. Sarah Lochlann Jain, "Malignant: How Cancer Becomes Us," Center for Science, Technology, Medicine, and Society at the University of California, Berkeley, February, 6, 2014, http://cstms.berkeley.edu/activ ities/live-stream/.

41. Michel Foucault, *The Birth of the Clinic: An Archaelogy of Medical Per-ception*, trans. A. M. Sheridan (London: Routledge, 1973), 3.

42. Foucault, *The Birth of the Clinic*.

43. Ehrenreich, "Welcome to Cancerland," 44.

6. DISPATCH FROM THE PALLIATIVE PRESENT

1. Emile Benveniste, "Subjectivity in Language," in *Problems in General Linguistics*, trans. Mary Elizabeth Meek (Coral Gables: University of Miami Press, 1971), 223–30 (226–27). For a sociographical inflection of the notion of tense, see Elizabeth Povinelli, *Economies of Abandon-ment: Social Belonging and Endurance in Late Liberalism* (Durham: Duke University Press, 2011), 11–14.

2. Roma Chatterji, "The Experience of Death in a Dutch Nursing Home: On Touching the Other," in *Living and Dying in the Contem-porary World: A Compendium*, ed. Veena Das and Clara Han (Berke-ley: University of California Press, 2015), 696–71 (706).

3. Chatterji, "The Experience of Death in a Dutch Nursing Home," 698.

4. Eric Cazdyn, *The Already Dead: The New Time of Politics, Culture, and Illness* (Durham: Duke University Press, 2012), 46. Canadian sociolo-gist Arthur W. Frank uses the expression "remission society" to describe patients that are "effectively well but could never be consid-ered cured." Arthur W. Frank, *The Wounded Storyteller: Body, Illness, and Ethics* (University of Chicago Press, 1995), 8.

5. René Descartes, *The Discourse on the Method* (1637), cited in Eric L. Krakauer, "'To Be Freed from the Infirmity of (the) Age': Subjectivity, Life-Sustaining Treatment, and Palliative Medicine," in *Subjectivity: Ethnographic Investigations* (Berkeley: University of California Press, 2007), 381–96 (387).

6. Joseph J. Fins and Nicholas D. Schiff, "In the Blink of the Mind's Eye," *Hasting Center Report* 40, no. 3 (January 2010): 21–23 (21). Martin M. Monti, Audrey Vanhaudenhuyse, Martin R. Coleman, Melanie Boly, John D. Pickard, Luaba Tshibanda, Adrian M. Owen, and Steven Laureys, "Willful Modulation of Brain Activity in Disorders of Consciousness," *New England Journal of Medicine* 362 (2010): 579–89.

7. J. Andrew Billings, Larry R. Churchill, and Richard Payne, "Severe Brain Injury and the Subjective Life," *Hasting Center Report* 40, no. 3 (January 2010): 17–21.

8. Max Weber, "Science as Vocation" ["*Wissenschaft als Beruf*"], in *From Max Weber: Essays in Sociology,* ed. H. H. Gerth and C. Wright Mills (New York: Oxford University Press, 1946), 129–56. For a philosophical genealogy of Weber's proposition, see Giorgio Agamben, *Infancy and History: The Destruction of Experience,* trans. Liz Heron (London: Verso, 1993), 23:

> Inasmuch as [the] goal [of traditional experience] was to advance the individual towards maturity—that is, an anticipation of death as the idea of an achieved totality of experience—it was something complete in itself, something it was possible to have, not only to undergo. But once experience was referred instead to the subject of science, which cannot reach maturity but can only increase its own knowledge, it becomes something incomplete . . . something it is possible only to *undergo*, never to *have*: nothing other, therefore, than the infinite process of knowledge.

9. Bichat, *Recherches physiologiques sur la vie et la mort,* 110. Cited in Agamben, *Remnants of Auschwitz,* 153 (translation modified).

10. Michel Foucault, *The Birth of the Clinic: An Archaelogy of Medical Perception,* trans. A. M. Sheridan (London: Routledge, 1973), 84. In his controversial "memoir" on the AIDS epidemic and Foucault's terminal

illness, *To the Friend Who Did Not Save My Life* (1990, for the French edition), Hervé Guibert relates an anecdote in which the author of *The Birth of the Clinic* asked to act as the benevolent "godfather" (*parrain*) for a daring project of boutique hospices. Foucault (under the pseudonym of Muzil) discredits the project—quite literally since a bank loan depended on his judgment: "that nursing home . . . shouldn't be a place where people go to die, but a place where they pretend to die. Everything there should be luxurious, with fancy paintings and soothing music, but it would all be just camouflage for the real mystery, because there'd be a little door hidden away in a corner of the clinic, perhaps behind one of those dreamily exotic pictures, and to the torpid melody of a hypodermic nirvana, you'd secretly slip behind the painting, and presto, you'd vanish, quite dead in the eyes of the world, since no one would see you reappear on the other side of the wall, in the alley, with no baggage, no name, no nothing, forced to invent a new identity for yourself." Hervé Guibert, *To the Friend Who Did Not Save My Life*, trans. Linda Coverdale (New York: Atheneum, 1991), 116–17. The anecdote might very well be apocryphal, but remains telling in the context of the palliative future of the birth (and the past) of the clinic: the spectacle of the terminal body is a disappearing act.

11. Nicholas A. Christakis, *Death Foretold: Prophecy and Prognosis in Medical Care* (Chicago: University of Chicago Press, 1999), 27. See also Jeffrey P. Bishop, *The Anticipatory Corpse: Medicine, Power, and the Care of the Dying* (Notre Dame: University of Notre Dame Press, 2011), 280–81.

12. Jean-Christophe Mino and Emmanuel Fournier, *Les Mots des derniers soins: La démarche palliative dans la médecine contemporaine* (Paris: Les Belles Lettres, 2008), 12.

13. Michel de Certeau, "The Unnameable," in *The Practice of Everyday Life*, 1:190–98 (190, translation modified).

14. Chatterji, "The Experience of Death in a Dutch Nursing Home," 698.

15. De Certeau, "The Unnameable," 196.

16. Greek inscription cited in Daniel Heller-Roazen, "Aglossostomography," *Parallax* 10, no. 1 (2004): 40–48 (46).

17. Jesper Svenbro, *Phrasikleia: Anthropologie de la lecture en Grèce ancienne* (Paris: Découverte, 1988), 37–38, translated and cited in Heller-Roazen, "Aglossostomography," 46.

18. Jean-Dominique Bauby, *The Diving Bell and the Butterfly* (New York: Knopf, 1997), 3–4 (translation modified). For the translation of "transport de cerveau," I rely on Tobias Watkins's 1809 translation of François Xavier Bichat, *Recherches physiologiques sur la vie et la mort* (Paris: Masson-Charpentier, 1852), 271: "In some paroxysms of acute fevers, the blood carried with violence to the brain, sometimes destroys its life. The patient has a delirium [*le malade a le transport*], as it is commonly called. If this delirium is raised to the highest degree, it proves mortal, and then the series of phenomena is the same as that in sudden deaths of which we have just spoken" (300).

19. Jean-Dominique Bauby, *Le Scaphandre et le papillon* (Paris: Laffont, 1997), 30.

20. On the collaborative dimension of disability narratives, see Thomas G. Couser, *Signifying Bodies: Disability in Contemporary Life Writing* (Ann Arbor: University of Michigan Press, 2009).

21. "The difference between kicking off (*crever*) and dying (*mourir*) is a *speech* that articulates, on the collapse of possessions and representations, the question: 'What does it mean to *be*?' . . . This is a speech that no longer says anything, that has nothing other than the loss out of which saying is formed. Between the machine that stops or kicks off, and the act of dying, there is the *possibility* of saying. The *possibility of dying* functions in this in-between space." De Certeau, "The Unnameable," 193 (emphasis in original).

22. Megan Craig, "Locked In," *Journal of Speculative Philosophy* 22, no. 3 (2008): 145–58 (151).

23. Dikaios Sakellariou, "Creating In/Abilities for Eating," *Somatosphere* (2015), http://somatosphere.net/2015/06/creating-inabilities-for-eating.html.

24. Paul Rabinow, *Essays on the Anthropology of Reason* (Princeton: Princeton University Press, 1996), 146.

25. Jean-Dominique Bauby, *The Diving Bell and the Butterfly* (New York: Knopf, 1997), 127.

26. Lawrence D. Kritzman, *The Fabulous Imagination: On Montaigne's Essays* (New York: Columbia University Press, 2009), 97.

27. Sarah Franklin and Margaret Lock, eds. *Remaking Life and Death: Toward an Anthropology of the Biosciences* (Santa Fe: School of American Research Press, 2003).

28. Jean-Luc Nancy, "L'Intrus," trans. Susan Hanson, *CR: The New Centennial Review* 2, no. 3 (2002): 1–14 (2–3).
29. Michel de Montaigne, "Of Practice," in *The Complete Essays of Montaigne*, trans. Donald M. Frame (Stanford: Stanford University Press, 1958): 267–75 (270, 267).
30. Montaigne, "Of Practice." 270.
31. Kritzman, *The Fabulous Imagination*, 87.

EPILOGUE

1. Roland Barthes, *Camera Lucida: Reflections on Photography*, trans. Richard Howard (New York: Hill and Wang, 1981), 10–11 (translation modified).
2. Barthes, *Camera Lucida*, 12 (translation modified). For an incisive analysis of the "mothering" properties of photography in *Camera Lucida*, see Elissa Marder, "Nothing to Say: Fragment on the Mother in the Age of Mechanical Reproduction," in *The Mother in the Age of Mechanical Reproduction: Psychoanalysis, Photography, Deconstruction* (New York: Fordham University Press, 2012), 149–59.
3. Michel Foucault, "The Lives of Infamous Men," in *Michel Foucault: Power, Truth, Strategy*, ed. Meaghan Morris and Paul Patton (Sydney: Feral, 1979), 76–91 (89).
4. Barthes, *Camera Lucida*, 15 (translation modified).
5. Ulrich Beck, "Climate for Change, or How to Create a Green Modernity?," *Theory, Culture, and Society* 27, no. 2–3 (2010): 254–66 (260).
6. Catherine Waldby and Robert Mitchell, *Tissue Economies: Blood, Organs, and Cell Lines in Late Capitalism* (Durham: Duke University Press, 2006), 120.
7. Sigmund Freud, "On Transience," in *The Standard Edition of the Complete Psychological Works of Sigmund Freud*, ed. and trans. James Strachey (London: Hogarth, 1957), 14:305–7 (305).

WORKS CITED

"12:31." Frank Schott photographie. Accessed on August 5, 2017. http://
www.frankschott.com/project/1231.

Agamben, Giorgio. *The Coming Community*. Trans. Michael Hardt. Minneapolis: University of Minnesota Press, 1993.

——. *Infancy and History: the Destruction of Experience*. Trans. Liz Heron.
London: Verso, 1993.

——. *Remnants of Auschwitz: The Witness and the Archive*. Trans. Daniel
Heller-Roazen. Brooklyn: Zone, 2002.

Agricola, Georgius. *De Re Metallica*. Trans. Herbert Clark Hoover and Lou
Henry Hoover. New York, Dover, 1950.

Alaimo, Stacy. "Material Memoirs: Science, Autobiography, and the Substantial Self." In *Bodily Natures: Science, Environment, and the Material
Self*, 85–112. Bloomington: Indiana University Press, 2010.

Andersen, Ross. "After Four Years, Checking Up on the Svalbard Global
Seed Vault." *Atlantic*, February 28, 2012.

Aniston, Jennifer. *Five*. U.S., 2011.

Appadurai, Arjun. "Grassroots Globalization and the Research Imagination." *Public Culture* 12, no. 1 (2000): 1–19.

Arasse, Daniel. "Le corps fictif de Sébastien et le coup d'œil d'Antonello."
In *Le corps et ses fictions*, 55–72. Ed. Claude Reichler. Geneva: Droz, 1983.

Archibald, Elizabeth. *Incest and the Medieval Imagination*. Oxford: Oxford
University Press, 2001.

Ausubel, Kenny. *Seeds of Change: The Living Treasure*. San Francisco: Harper
Collins, 1994.

Barthes, Roland. *Camera Lucida: Reflections on Photography*. Trans. Richard Howard. New York: Hill and Wang, 1981.

——. "The World as Object." In *Critical Essays*, 3–12. Trans. Richard Howard. Evanston: Northwestern University Press, 2000.

Bauby, Jean-Dominique. *The Diving Bell and the Butterfly*. New York: Knopf, 1997.

——. *Le Scaphandre et le papillon*. Paris: Laffont, 1997.

Beck, Ulrich. "Climate for Change, or How to Create a Green Modernity?" *Theory, Culture, and Society* 27, nos. 2–3 (2010): 254–66.

Belling, Catherine. "Narrating Oncogenesis: The Problem of Telling When Cancer Begins." *Narrative* 18, no. 2 (2010): 229–47.

Belting, Hans. *An Anthropology of Images: Picture, Medium, Body*. Trans. Thomas Dunlap. Princeton: Princeton University Press, 2011.

Benjamin, Walter. "The Storyteller: Reflections on the Works of Nikolai Leskov." In *Illuminations*, 83–110. Ed. Hannah Arendt. Trans. Harry Zohn. New York: Schocken, 2007.

Benveniste, Emile. "Subjectivity in Language." In *Problems in General Linguistics*, 223–30. Trans. Mary Elizabeth Meek. Coral Gables: University of Miami Press, 1971.

Berger, Harry. *Caterpillage: Reflections on Seventeenth-Century Dutch Still Life Painting*. New York: Fordham University Press, 2011.

Berger Hochstrasser, Julie. *Still Life and Trade in the Dutch Golden Age*. New Haven: Yale University Press, 2007.

Berlant, Lauren. *Cruel Optimism*. Durham: Duke University Press, 2011.

Bichat, Marie François Xavier. *Physiological Researches Upon Life and Death*. Trans. Tobias Watkins. Philadelphia: Smith and Maxwell, 1809.

——. *Recherches physiologiques sur la vie et la mort*. Paris: Masson-Charpentier, 1852.

Bidart, Frank. "The Second Hour of the Night." *Threepenny Review* 69 (Spring 1997): 6–13.

Billings, J. Andrew, Larry R. Churchill, and Richard Payne. "Severe Brain Injury and the Subjective Life." *Hasting Center Report* 40, no. 3 (January 2010): 17–21.

Biotechnology Law Report 8, no. 6 (November-December 1989): 489–90.

Bishop, Jeffrey P. *The Anticipatory Corpse: Medicine, Power, and the Care of the Dying*. Notre Dame: University of Notre Dame Press, 2011.

Boyle, James. *Shamans, Software, and Spleens: Law and the Construction of the Information Society*. Cambridge: Harvard University Press, 1997.

Brophy, Philip. "Horrality—the Textuality of Contemporary Horror Films." In *The Horror Reader*, 276–84. Ed. Ken Gelder. London: Routledge, 2000.

Bruening, G., and J. M. Lyons. "The Case of the FLAVR SAVR Tomato." *California Agriculture* 54, no. 4 (July-August 2000): 6–7.

Brandt, Steward. "The Case for De-Extinction: Why We Should Bring Back the Woolly Mammoth." *Environment* 360 (January13, 2014). Accessed July 27, 2017. http://e360.yale.edu/feature/the_case_for_de-extinction_why_we_should_bring_back_the_woolly_mammoth/2721.

Brekk, Lars Peder. "Frozen Seeds in a Frozen Mountain—Feeding a Warming World!" Accessed July 27, 2017. http://www.regjeringen.no/en/dep/lmd/whats-new/Speeches-and-articles/speeches-and-articles-by-the-minister/speeches-and-articles-/one-year-anniversary-seminar-of-the-sval/one-year-anniversary-seminar-of-the-sval.html?id=547254.

Brooks, Peter. *Reading for the Plot: Design and Intention in Narrative*. Cambridge: Harvard University Press, 1992.

Bryson, M. K., and Chase Joynt. "Under the Skin: Imag(in)ing Medicine's Queer Pedagogies as Moving Pictures." *No More Potlucks* 29 (2013). Accessed July 15, 2017. http://nomorepotlucks.org/site/under-the-skin-imagining-medicinesqueer-pedagogies-as-moving-pictures-chase-joynt-m-k-bryson.

Bryson, Norman. *Looking at the Overlooked: Four Essays on Still Life Painting*. Cambridge: Harvard University Press, 1990.

Bynum, Caroline Walker. "Material Continuity, Personal Survival, and the Resurrection of the Body: A Scholastic Discussion in Its Medieval and Modern Contexts." In *Fragmentation and Redemption: Essays on Gender and the Human Body in Medieval Religion*, 239–97. New York: Zone, 1991.

Canguilhem, Georges. "Diseases." In *Writings on Medicine*, 34–41. Trans. Stefanos Geroulanos and Todd Meyers. New York: Fordham University Press, 2012.

——. *The Normal and the Pathological*. Trans. Carolyn R. Fawcett in collaboration with Robert S. Cohen. New York: Zone, 1989.

Carlino, Andrea. "The Book, the Body, the Scalpel: Six Engraved Title Pages for Anatomical Treatises of the First Half of the Sixteenth Century." *RES: Anthropology and Aesthetics* 16 (Autumn 1988): 33–50.

Cartwright, Lisa. "The Visible Man: The Man Criminal Subject as Biomedical Norm." In *Processed Lives: Gender and Technology in Everyday*

Life, 86–100. Ed. Melodie Calvert and Jennifer Terry. New York: Routledge, 1997.

Cazdyn, Eric. *The Already Dead: The New Time of Politics, Culture, and Illness*. Durham: Duke University Press, 2012.

Chambers, Tod. *The Fiction of Bioethics: Cases as Literary Texts*. New York: Routledge, 1999.

Charon, Rita. *Narrative Medicine: Honoring the Stories of Illness*. New York: Oxford University Press, 2008.

Chatterji, Roma. "The Experience of Death in a Dutch Nursing Home: On Touching the Other." In *Living and Dying in the Contemporary World: A Compendium*, 696–711. Ed. Veena Das and Clara Han. Berkeley: University of California Press, 2015.

Christakis, Nicholas A. *Death Foretold: Prophecy and Prognosis in Medical Care*. Chicago: University of Chicago Press, 1999.

Craig, Megan. "Locked In." *Journal of Speculative Philosophy* 22, no. 3 (2008): 145–58.

Certeau, Michel de. *The Practice of Everyday Life*, vol. 1. Trans. Steven Rendall. Berkeley: University of California Press, 1984.

——. *The Writing of History*. Trans. Tom Conley. New York: Columbia University Press, 1988.

Couser, Thomas G. *Signifying Bodies: Disability in Contemporary Life Writing*. Ann Arbor: University of Michigan Press, 2009.

Deleuze, Gilles. *Foucault*. Trans. Seán Hand. London: Continuum, 1999.

Derrida, Jacques. "Of an Apocalyptic Tone Newly Adopted in Philosophy." In *Derrida and Negative Theology*, 25–71. Ed. Harold G. Coward and Toby Foshay. Albany: SUNY Press, 1992.

Didi-Huberman, Georges. "Ex-Voto: Image, Organ, Time." *L'Esprit Créateur* 47, no. 3 (2007): 7–16.

Diedrich, Lisa. *Treatments: Language, Politics, and the Culture of Illness*. Minneapolis: University of Minnesota Press, 2007.

Duden, Barbara. "Die Verkrebsung: Kränkende Diagnostik durch Krebsprävention." In *Die Gene im Kopf—der Fötus im Bauch. Historiches zum Frauenkörper*, 166–86. Hanover: Offizin, 2002.

——. *Disembodying Women: Perspectives on Pregnancy and the Unborn*. Trans. Lee Hoinacki. Cambridge: Harvard University Press, 1993.

——. "A Historian's 'Biology': On the Traces of the Body in a Technogenic World." *Historein* 3 (2001): 89–102.

Edelman, Lee. *No Future: Queer Theory and the Death Drive.* Durham: Duke University Press, 2005.

Ehrenreich, Barbara. "Welcome to Cancerland: A Mammogram Leads to a Cult of Pink Kitsch." *Harper's,* November 2001, 43–53.

Feder, Toni. "DOE Opens WIPP for Nuclear Waste Burial." *Physics Today* 52, no. 5 (1999): 59.

Ferguson, Frances. "The Nuclear Sublime," *diacritics* 14, no. 2 (Summer 1984): 4–10.

Fins, Joseph J., and Nicholas D. Schiff. "In the Blink of the Mind's Eye." *Hasting Center Report* 40, no. 3 (January 2010): 21–23.

Flitterman-Lewis, Sandy. "From *Déesse* to *Idée: Cleo from 5 to 7.*" In *To Desire Differently: Feminism and the French Cinema,* 268–84. Rev. ed. New York: Columbia University Press, 1996.

Foucault, Michel. *The Birth of the Clinic: An Archaelogy of Medical Perception.* Trans. A. M. Sheridan. London: Routledge, 1973.

——. *The Hermeneutics of the Subject: Lectures at the Collège de France, 1981–1982.* Trans. Graham Burchell. New York: Picador, 2005.

——. *History of Sexuality.* vol. 1: *An Introduction.* Trans. Robert Hurley. New York: Pantheon, 1980.

——. "The Life of Infamous Men." In *Michel Foucault: Power, Truth, Strategy.* Ed. Meaghan Morris and Paul Patton, 76–91. Sydney: Feral, 1979.

François, Anne-Lise. "Flower Fisting." *Postmodern Culture* 22, no. 1 (2011).

——. "'O Happy Living Things' Frankenfoods and the Bounds of Wordsworthian Natural Piety." *diacritics* 33, no. 2 (2005): 42–70.

Frank, Arthur W. *The Wounded Storyteller: Body, Illness, and Ethics.* Chicago: University of Chicago Press, 1995.

Franklin, Sarah, and Margaret Lock, eds. *Remaking Life and Death: Toward an Anthropology of the Biosciences.* Santa Fe: School of American Research Press, 2003.

Franklin, Sarah, Celia Lury, and Jackie Stacey. *Global Nature, Global Culture.* London: SAGE, 2000.

Freud, Sigmund. "On Transience." In *The Standard Edition of the Complete Psychological Works of Sigmund Freud.* Ed. and trans. James Strachey, 14:305–7. London: Hogarth, 1957.

Gombrich, Ernst H. "Tradition and Expression in Western Still-life." *Burlington Magazine* 103 (1961): 175–80.

Grootenboer, Hanneke. *The Rhetoric of Perspective: Realism and Illusionism in Seventeenth- Century Dutch Still-Life Painting.* Chicago: University of Chicago Press, 2005.

Guibert, Hervé. *To the Friend Who Did Not Save My Life.* Trans. Linda Coverdale. New York: Atheneum, 1991.

Gumbrecht, Hans Ulrich. "Shall We Continue to Write Histories of Literature?" *New Literary History* 39, no. 3 (2008): 519–32.

Haraway, Donna. *Modest_Witness@Second_Millennium.FemaleMan_Meets _OncoMouse: Feminism and Technoscience.* New York: Routledge, 1997.

Harpham, Geoffrey Galt. "The Humanities in America." Keynote address for the conference "The Humanities Into the Twenty-first Century." University of Copenhagen. Accessed on March 15, 2017. http://hum21 .ku.dk/humanities_in_a_new_millenium/geoffrey_galt_harpham/.

——. "The Humanities' Value." *Chronicle of Higher Education* 55, no. 28 (March 20, 2009): B6.

Heller-Roazen, Daniel. "Aglossostomography." *Parallax* 10, no. 1 (2004): 40–48.

Herring, Scott. *The Hoarders: Material Deviance in Modern American Culture.* Chicago: University of Chicago Press, 2014.

Hird, Myra. "Landscapes of Terminal Capitalism, Aporias of Responsibility: Lifeworlds Inherited, Inhabited and Bequeathed." "Conference Anthropocene Feminism." Center for Twenty-First-Century Studies, University of Wisconsin-Milwaukee. April 2014.

——. "Waste, Landfills, and an Environmental Ethic of Vulnerability." *Ethics and the Environment* 18, no. 1 (2013): 105–24.

Holloway, Karla F. C. *Private Bodies, Public Texts: Race, Gender, and Cultural Bioethics.* Durham: Duke University Press, 2011.

"How Will Future Generations Be Warned?" Accessed on March 25, 2016. http://www.wipp.energy.gov/fctshts/PICs.pdf.

Huffer, Lynne. *Mad for Foucault: Rethinking the Foundations of Queer Theory.* New York: Columbia University Press, 2009.

Jeanneret, Michel. "Les paroles dégelées (Rabelais, *Quart Livre*, 48–65)." *Littérature* 17 (1975): 14–30.

Jurecic, Ann. *Illness as Narrative.* Pittsburgh: University of Pittsburgh Press, 2012.

Kafer, Alison. *Feminist, Queer, Crip.* Bloomington: Indiana University Press, 2013.

Kafka, Franz. "In the Penal Colony." Trans. Ian Johnston. Accessed on August 6, 2017. http://records.viu.ca/~johnstoi/kafka/inthepenalcol ony.htm.

Kahn, Hermann. *On Thermonuclear War*. Princeton: Princeton University Press, 1960.

Kamuf, Peggy. "Life in Storage: Of Capitalism and A&E's 'Storage Wars.'" *Los Angeles Review of Books*, February 4, 2012.

Kantorowicz, Ernst. *The King's Two Bodies: A Study in Medieval Political Theology*. Princeton: Princeton University Press, 1957.

King, Allan, dir. *Dying at Grace*. 2003; New York: Criterion Collection, 2010. DVD.

Klose, Alexander. *The Container Principle: How a Box Changes the Way We Think*. Trans. Charles Marcrum. Cambridge: MIT Press, 2015.

Knorr Cetina, Karin. "The Rise of a Culture of Life." *EMBO Reports* 6 (2005): 76–80.

Krakauer, Eric L. "'To Be Freed From the Infirmity of (the) Age': Subjectivity, Life-Sustaining Treatment, and Palliative Medicine." In *Subjectivity: Ethnographic Investigations*, 381–396. Ed. João Biehl, Byron J. Good, and Arthur Kleinman. Berkeley: University of California Press, 2007.

Kritzman, Lawrence D. *The Fabulous Imagination: On Montaigne's Essays*. New York: Columbia University Press, 2009.

Lampic, C., A. Skoog Svanber, P. Karlström, and T. Tydén. "Fertility Awareness, Intentions Concerning Childbearing, and Attitudes Towards Parenthood Among Female and Male Academics." *Human Reproduction* 21, no. 2 (February 2006): 558–64.

Landecker, Hannah. *Culturing Life: How Cells Became Technologies*. Cambridge: Harvard University Press, 2010.

——. "Food as Exposure: Nutritional Epigenetics and the New Metabolism." *BioSocieties* 6 (2011): 167–94.

Leathers Kuntz, Marion. "Rabelais, Postel et utopie." In *Travaux d'Humanisme et Renaissance*, no. 321, *Études Rabelaisiennes*, vol. 33, *Rabelais pour le XXIe siècle. Actes du colloque du Centre d'Etudes Supérieures de la Renaissance (Chinon-Tours, 1994)*, 55–78. Ed. Michel Simonin. Genève: Droz, 1998.

Leigh, Julia. *The Hunter*. New York: Four Walls Eight Windows, 1999.

Lewis, Philip. *Seeing Through the Mother Goose Tales: Visual Turns in the Writings of Charles Perrault*. Stanford: Stanford University Press, 1996.

Liao, Matthew, Anders Sandberg, and Rebecca Roache. "Human Engineering and Climate Change." *Ethics, Policy and Environment* 15, no. 2 (June 2012): 206–21.

Lochlann Jain, Sarah. "Living in Prognosis: Toward an Elegiac Politics." *Representations* 98 (Spring 2007): 77–92.

——. *Malignant: How Cancer Becomes Us.* Berkeley: University of California Press, 2013.

——. "Malignant: How Cancer Becomes Us." Talk at the Center for Science, Technology, Medicine, and Society. University of California, Berkeley. February 6, 2014. Accessed August 4, 2017. http://cstms.berkeley.edu/activities/live-stream/.

Löwy, Ilana. *Preventive Strikes: Women, Precancer, and Prophylactic Surgery.* Baltimore: Johns Hopkins University Press, 2010.

Lyotard, Jean-François. "*Oikos.*" In *Political Writings*, 96–107. Trans. Bill Readings and Kevin Geiman. Minneapolis: University of Minnesota Press, 1993.

——. *Postmodern Fables.* Trans George Van Den Abbeele. Minneapolis: University of Minnesota Press, 1997.

——. "Prescription." Trans. Christopher Fynsk. *L'Esprit Créateur* 31, no. 1 (1991): 15–32.

Marder, Elissa. *The Mother in the Age of Mechanical Reproduction: Psychoanalysis, Photography, Deconstruction.* New York: Fordham University Press, 2012.

Marin, Louis. *On Representation.* Trans. Catherine Porter. Stanford: Stanford University Press, 2002.

Marx, Karl. *Capital. A Critique of Political Economy*, vol. 1. Ed. Ernest Mandel Trans. Ben Fowkes. Harmondsworth: Penguin, 1982.

Masco, Joseph. "Mutant Ecologies: Radioactive Life in Post–Cold War New Mexico." *Cultural Anthropology* 19, no. 4 (2004): 517–50.

——. *The Nuclear Borderlands: The Manhattan Project in Post–Cold War New Mexico.* Princeton: Princeton University Press, 2006.

Madsen, Michael, dir. *Into Eternity a Film for the Future.* 2010; San Francisco, CA: Distributed by Video Project, 2010. DVD.

Masheck, Joseph. "Alberti's 'Window': Art-Historiographic Notes on an Antimodernist Misprision." *Art Journal* 50, no. 1 (Spring 1991): 34–41.

Mauss, Marcel. "A Category of the Human Mind: The Notion of Person; The Notion of Self." In *The Category of the Person: Anthropology,*

Philosophy, History, 1–25. Ed. Michael Carrithers, Steven Collins, and Steven Lukes. Cambridge: Cambridge University Press, 1985.

Medovoi, Leerom. "A Contribution to the Critique of Political Ecology: Sustainability as Disavowal." *New Formations* 69 (2010): 129–43.

Menninghaus, Winfried. *In Praise of Nonsense: Kant and Bluebeard*. Trans. Henry Pickford. Stanford: Stanford University Press, 1999.

Meyers, Todd. *The Clinic and Elsewhere: Addiction, Adolescents, and the After-life of Therapy*. Seattle: University of Washington Press, 2013.

Miah, Andy. "Justifying Human Enhancement: The Accumulation of Biocultural Capital." In *The Transhumanist Reader: Classical and Contemporary Essays on the Science, Technology, and Philosophy of the Human Future*, 291–301. Ed Max More and Natasha Vita-More. Malden, MA: Wiley-Blackwell, 2013.

Mino, Jean-Christophe, and Emmanuel Fournier. *Les Mots des derniers soins: La démarche palliative dans la médecine contemporaine*. Paris: Les Belles Lettres, 2008.

Mitchell, Richard G. *Dancing at Armageddon: Survivalism and Chaos in Modern Times*. Chicago: Chicago University Press, 2002.

Montaigne, Michel de. *The Complete Essays of Montaigne*. Trans. Donald M. Frame. Stanford: Stanford University Press, 1958.

——. *Essais*. Ed. Villey-Saulnier. Paris: Presses Universitaires de France, 2004.

Monti, Martin M., Audrey Vanhaudenhuyse, Martin R. Coleman, Melanie Boly, John D. Pickard, Luaba Tshibanda, Adrian M. Owen, and Steven Laureys. "Willful Modulation of Brain Activity in Disorders of Consciousness." *New England Journal of Medicine* 362 (2010): 579–89.

Montross, Christine. *Body of Work: Meditations on Mortality from the Human Anatomy Lab*. New York: Penguin, 2008.

Moore, Marlon Rachquel. "Opposed to the Being of Henrietta: Bioslavery, Pop Culture, and the Third Life of HeLa cells." *Medical Humanities* 43 (2017): 55–61.

Morgan, Lynn. "'Properly Disposed Of': A History of Embryo Disposal and the Changing of Claims on Fetal Remains." *Medical Anthropology* 21, nos. 3–4 (2002): 247–74.

Nachtigall, R. D., K. MacDougall, J. Harrington, J. Duff, G. Lee, and M. Becker. "How Couples Who Have Undergone In Vitro Fertilization

Decide What to Do with Surplus Frozen Embryos." *Fertility and Sterility* 92, no. 6 (December 2009): 2094–96.

Nancy, Jean-Luc. "L'Intrus." Trans. Susan Hanson. *CR: The New Centennial Review* 2, no. 3 (2002): 1–14.

Nixon, Rob. *Slow Violence and the Environmentalism of the Poor.* Cambridge: Harvard University Press, 2011.

Palladino, Paolo. *Plants, Patients, and the Historian: (Re)membering in the Age of Genetic Engineering.* Manchester: Manchester University Press, 2002.

Paulson, William. *Literary Culture in a World Transformed. A Future for the Humanities.* Ithaca: Cornell University Press, 2001.

Perrault, Charles. *Histoires ou Contes du Temps Passé. Avec des Moralitez.* Paris: Barbin, 1697.

Pfeffer Merrill, Jacqueline. "Embryos in Limbo." *New Atlantis* 24 (Spring 2009): 18–28.

Pimm, Stuart. "Opinion: The Case Against Species Revival," *National Geographic*, March 12, 2013.

Pine, Jason. "Economy of Speed: The New Narco-Capitalism." *Public Culture* 19, no. 2 (2007): 357–66.

Pinkus, Karen. "Carbon Management: A Gift of Time?" *Oxford Literary Review* 32, no. 1 (2010): 51–70.

——. *Fuel: A Speculative Dictionary.* Minneapolis: University of Minnesota Press, 2016.

——. "The Risks of Sustainability." In *Criticism, Crisis, and Contemporary Narrative: Textual Horizons in an Age of Global Risk*, 62–80. Ed. Paul Crosthwaite. London: Routledge, 2011.

Pinkus, Karen, and Cameron Tonkinwise. "Want Not: A Dialogue on Sustainability with Images." *world picture* 5 (Spring 2011): 1–11. Accessed on April 5, 2017. http://www.worldpicturejournal.com/WP_5/PDFs/Pinkus_Tonkinwise.pdf.

Povinelli, Elizabeth. "After the Last Man: Images and Ethics of Becoming Otherwise," *e-flux* 35 (May 2012). Acessed on August 5, 2017. http://www.e-flux.com/journal/35/68380/after-the-last-man-images-and-ethics-of-becoming-otherwise/.

——. *Economies of Abandonment: Social Belonging and Endurance in Late Liberalism.* Durham: Duke University Press, 2011.

——. "New Media/Other Worlds?" Lecture at ICI Berlin. October 2011.

"Object Breast Cancer." *Studio 360* (2012). Accessed on August 4, 2017. https://www.wnyc.org/radio/#/ondemand/227070.

Ong, Aihwa, and Stephen J. Collier. *Global Assemblages: Technology, Politics, and Ethics as Anthropological Problems*. Malden, MA: Blackwell, 2005.

Ovid, *Metamorphoses*, X. Trans. Anthony S. Kline. Accessed on August 5, 2017. http://ovid.lib.virginia.edu/trans/Metamorph10.htm#484521426.

Rabelais, François. *The Fourth Book*. In *The Complete Works of François Rabelais*. Trans. Donald M. Frame. Berkeley: University of California Press, 1991.

——. *Œuvres complètes*. Ed. Mireille Huchon, with François Moreau. Paris: Gallimard-NRF, 1994.

Rabinow, Paul. *Anthropos Today: Reflections on Modern Equipment*. Princeton: Princeton University Press, 2003.

——. *Essays on the Anthropology of Reason*. Princeton: Princeton University Press, 1996.

Readings, Bill. *The University in Ruins*. Cambridge: Harvard University Press, 1997.

——. "University Without Culture." *New Literary History* 26, no. 3 (1995): 465–92.

Roy, Susan. *Bomboozled: How the U.S. Government Misled Itself and Its People Into Believing They Could Survive a Nuclear Attack*. New York: Pointed Leaf, 2011.

Sahai, Suman. "The Bogus Debate of Bioethics." *Biotechnology and Development Monitor* 30 (March 1997).

Sakellariou, Dikaios. "Creating In/Abilities for Eating." *Somatosphere* (2015). Accessed on March 17, 2017. http://somatosphere.net/2015/06/creating-inabilities-for-eating.html.

Scarry, Elaine. *Thinking in an Emergency*. New York: Norton, 2011.

Schnabel, Julian, dir. *Le scaphandre et le papillon*. 2007; Burbank, CA: Touchstone Home Entertainment, 2008. DVD.

Seabrook, John. "Sowing for Apocalypse: The Quest for a Global Seed Bank." *New Yorker*, August 27, 2007.

Sebeok, Thomas A. *Communication Measures to Bridge Ten Millennia*. Columbus, OH: Office of Nuclear Waste Isolation, Battelle Memorial Institute, 1984.

Serres, Michel. *Malfeasance: Appropriation Through Pollution?* Trans. Anne-Marie Feenberg-Dibon. Stanford: Stanford University Press, 2010.

Skloot, Rebecca. *The Immortal Life of Henrietta Lacks.* New York: Crown, 2010.

Skloot, Rebecca. "The Immortal Life of Henrietta Lacks, the Sequel." *New York Times*, March 23, 2013.

Smith, Adam. *An Inquiry Into the Nature and Causes of the Wealth of Nations* (1776). Ed. Edwin Cannan. London: Methuen, 1904. Accessed on August 4, 2017. http://www.econlib.org/library/Smith/smWN.html.

Spillers, Hortense J. "'Mama's Baby, Papa's Maybe: An American Grammar Book." *diacritics* 17, no. 2 (1987): 64–81.

Stacey, Jackie. *Teratologies: A Cultural Study of Cancer.* London: Routledge, 1997.

Sterling, Charles. *Still Life Painting: From Antiquity to the Twentieth Century.* New York : Harper and Row, 1981.

Stewart, Kathleen. *Ordinary Affects.* Durham: Duke University Press, 2007.

Stoekl, Allan. "'After the Sublime,' After the Apocalypse: Two Versions of Sustainability in Light of Climate Change." *diacritics* 41, no. 3 (2013): 40–57.

——. "Agnès Varda and the Limits of Gleaning." *world picture* 5 (Spring 2011). Accessed March 3, 2017. http://www.worldpicturejournal.com/WP_5/Stoekl.html.

——. *Bataille's Peak: Energy, Religion, and Postsustainability.* Minneapolis: University of Minnesota Press, 2007.

——. "Gift, Design and Gleaning." *Design Philosophy Papers* 7, no. 1 (2009): 7–17.

Svalbard Global Seed Vault. Norwegian Ministry of Agriculture and Food website. Accessed on August 5, 2017. http://www.regjeringen.no/en/dep/lmd/campain/svalbard-global-seed-vault.html?id=462220.

Svenbro, Jesper. *Phrasikleia: Anthropologie de la lecture en Grèce ancienne.* Paris: La Découverte, 1988.

Svendsen, Mette. "Articulating Potentiality: Notes on the Delineation of the Blank Figure in Human Embryonic Stem Cell Research." *Cultural Anthropology* 26, no. 3 (2011): 414–37.

Tonkinwise, Cameron. "Practicing Sustainability by Design: Global Warming Politics in a Post-awareness World." *Scapes* 6 (Fall 2007): 4–12.

———. "Sustainability Is Not a Humanism: Review Essay on Allan Stoekl's *Bataille's Peak*." *Design Philosophy Papers* 7, no. 1 (2009): 39–48.

"UK Pavillion Shangai Expo 2010." Heatherwick Studio. Accessed August 3, 2017. http://www.heatherwick.com/uk-pavilion/.

United Nations. *Our Common Future: The World Commission on Environment and Development*. Oxford: Oxford University Press, 1987.

Van Dijck, Jose. *The Transparent Body: A Cultural Analysis of Medical Imaging*. Seattle: University of Washington Press, 2005.

Van Wyck, Peter. *Signs of Danger; Waste, Trauma, and Nuclear Threat*. Minneapolis: University of Minnesota Press, 2005.

Varda, Agnès, dir. *The Gleaners and I*. 2000; New York: Zeitgeist Video, 2002. DVD.

Vernant, Jean-Pierre. "At Man's Table: Hesiod's Foundation Myth of Sacrifice." In *The Cuisine of Sacrifice Among the Greeks*, 21–86. Ed. Marcel Detienne and Jean-Pierre Vernant. Trans. Paula Wissing. Chicago: Chicago University Press, 1989.

"Visible Human Project" (The). U.S. National Library of Medicine. Accessed on August 4, 2017. http://www.nlm.nih.gov/research/visible /visible_human.html.

Wald, Priscilla. "What's in a Cell?: John Moore's Spleen and the Language of Bioslavery." *New Literary History* 36, no. 2 (2005): 205–25.

Waldby, Catherine. "Stem Cells, Tissue Cultures and the Production of Biovalue." *Health: An Interdisciplinary Journal for the Social Study of Health, Illness, and Medicine* 6, no. 3 (2002): 305–23.

———. *The Visible Human Project: Informatic Bodies and Posthuman Medicine*. London: Routledge, 2000.

Waldby, Catherine, and Robert Mitchell. *Tissue Economies: Blood, Organs, and Cell Lines in Late Capitalism*. Durham: Duke University Press, 2006.

Weber, Max. "Science as Vocation" ["*Wissenschaft als Beruf*"]. In *From Max Weber: Essays in Sociology*, 129–56. Ed. and trans. H. H. Gerth and C. Wright Mills. New York: Oxford University Press, 1946.

Weber, Samuel. *Institution and Interpretation*. Expanded ed. Stanford: Stanford University Press, 2001.

Weheliye, Alexander G. *Habeas Viscus: Racializing Assemblages, Biopolitics, and Black Feminist Theories of the Human*. Durham: Duke University Press, 2014.

Wilson, Robin. "Timing Is Everything: Academe's Annual Baby Boom." *Chronicle of Higher Education*, June 25, 1999.

Winnicott, Donald. *Playing and Reality*. London: Routledge, 1991.

Wood, Gillen D'Arcy. "What Is Sustainability?" *American Literary History* 24, no. 1 (2012): 1–15 (10).

Woodward, Kathleen. "Statistical Panic." *differences: A Journal of Feminist Cultural Studies* 11, no. 2 (1999). 177–203.

INDEX

Svalbard Global Seed Vault
  project (Svalbard project), *3*,
  13; biodiversity and, 2, 25;
  doomsday event and, 2, 5; as
  form-giving event, 10–11;
  Fowler on, 37–38, 115*n*7;
  frozen words compared to, 10;
  location and contents of, 1; Seed
  Cathedral and, 2, *4*; still life and,
  xvi, 25; as sustainability model, 5
Svalbard project. *See* Svalbard
  Global Seed Vault project
Svenbro, Jesper, 96
Swayze, Jay, 123*n*31

Tasmanian tiger. *See* Thylacine
Terminal capitalism, 47
Terminal humanities, 122*n*23
Terminal Illness. *See* Palliative care
Thylacine (Tasmanian tiger), 30–31
Tissue culture: portraiture of,
  60; storage and burial of, 49;
  value of, 68. *See also* Cellular
  immortality; HeLa cells
Tonkinwise, Cameron, xiii
Toxic waste, x; Hird on, 30; nuclear
  waste management, xvii; Sebeok
  on transuranic waste burial,
  36–37, 38; terminal humanities
  and, 122*n*23

The University in Ruins (Readings),
  40

Valverde de Amusco, Juan, *65*, 66
Vanitas, 12, 22–23
Van Wyck, Peter, 35
Varda, Agnès, 48, 81–83
"Vergänglichkeit" ("On Transience")
  (Freud), 111–12
Vernant, Jean-Pierre, 31–32
Verpleeghuis (Dutch nursing home),
  90
VHP. *See* Visible Human Project
Viability: Canguilhem on, viii, x,
  104; perishability compared
  to, x
Visible Human Project (VHP),
  62; Jernigan in, 61, 63, 64;
  recapitulation effect for,
  63–64

Waldby, Catherine, 61, 63, 67
Waste Isolation Pilot Plant
  (WIPP): continuity in time
  of, 35; symbolic leak of, 35;
  transmission logic for, 38
Weber, Max, 92, 131*n*8
Weheliye, Alexander, 53
Winnicott, Donald, xv–xvi,
  114*n*19
WIPP. *See* Waste Isolation Pilot
  Plant
Wood, Gillen D'Arcy, xiii–xiv
Woodward, Kathleen, 79–81
Words of Terminal Care, The. *See*
  Mots des derniers soins, Les

GPSR Authorized Representative: Easy Access System Europe, Mustamäe tee 50, 10621 Tallinn, Estonia, gpsr.requests@easproject.com

www.ingramcontent.com/pod-product-compliance
Lightning Source LLC
Chambersburg PA
CBHW032138020426
42334CB00016B/1212